Dynamics of
Biological Invasions

Dynamics of
Biological Invasions

Rob Hengeveld

Research Institute for Nature Management
The Netherlands

London
New York

CHAPMAN AND HALL

First published in 1989 by
Chapman and Hall Ltd
11 New Fetter Lane, London EC4P 4EE

Published in the USA by
Chapman and Hall Ltd
29 West 35th Street, New York NY 10001

Typeset in 10/12 pt Parlament by
Leaper & Gard Ltd, Bristol
Printed in Great Britain by
St Edmundsbury Press, Bury St Edmunds, Suffolk

ISBN 0 412 31470 3

British Library Cataloguing in Publication Data

Hengeveld, Rob
    Dynamics of biological invasions.
    1.   Ecosystems.   Effects on alien plants   2.
    Ecosystems.   Effects on alien animals
    I.   Title
    574.5

ISBN 0-412-31470-3

Library of Congress Cataloging in Publication Data

Applied for

Contents

Preface   ix
Acknowledgements   x
Introduction   xi

PART ONE THEORY   **1**

1 Perspectives of Biological Invasions   **3**
   1.1   The ecological perspective   3
   1.2   The genetical perspective   4
   1.3   The epidemiological perspective   5
   1.4   The biogeographical perspective   5
   1.5   The mathematical perspective   6
   1.6   Terminology   6
   1.7   Conclusions   7

2 Examples of invasions   **9**
   2.1   Holocene tree invasions   9
   2.2   A genetical wave of advance in man   11
   2.3   The red deer and the thar in New Zealand 14
   2.4   Cholera in America and measles in Iceland 18
   2.5   Recurrent population waves   21
   2.6   Conclusion   24

3 Measures of areal expansion   **26**
   3.1   Relative and absolute radial increase   26
   3.2   Areal circumference   27
   3.3   The square root of the area occupied   28
   3.4   Conclusions   29

4 Population growth   **30**
   4.1   Exponential growth   30

4.2 Microepidemics 32
4.3 Logistic growth 34
4.4 Logistic growth on different spatial scales 37
4.5 An example: the European starling in North America 39
4.6 Conclusions 42

5 Diffusion **44**
5.1 Neighbourhood diffusion 46
5.2 Stratified diffusion 48
5.3 Some probability distributions 53
5.4 The dispersion probability field 57
5.5 Diffusion and population growth: the advancing-wave model 60
5.6 General transport models 62
5.7 An example: the muskrat in Europe 65
5.8 Conclusions 73

PART TWO   APPLICATIONS AND INTERPRETATION **75**

6 Parameter estimation and ecological boundary conditions **77**
6.1 Parameter estimation 77
6.2 Environmental limitation 80
6.3 Two examples 80
6.4 Conclusions 84

7 Simulating biological invasions **85**
7.1 The immigration of Neolithic farmers into Europe 86
7.2 The spread of stripe rust in wheat 88
7.3 Conclusions 90

8 Birds invading Europe and America **92**
8.1 The European invasion of the collared dove 92
8.2 Four other European invaders 103
8.3 Two American invaders 112
8.4 Conclusions 114

9 The stochastic structure of the wave front of rabies in central Europe **116**
9.1 The progression of the wave front 117
9.2 Foci within the wave front 119
9.3 Temporal structure of the wave front 121
9.4 Conclusions 124

10 Interpreting biological invasions   **126**
    10.1   The balance-of-nature paradigm   126
    10.2   Invaders and the demography of species and communities   127
    10.3   The stationarity of species and community processes   129
    10.4   A non-equilibrium approach   130
    10.5   Invasions as anomalies of the balance-of-nature paradigm   131
    10.6   The historical paradigm of biogeography   131
    10.7   Invasions as anomalies of the historical paradigm   133
    10.8   Developments towards a dynamic paradigm   133
    10.9   Invasions and measures of control and conservation   137
    10.10 Conclusions   139

11 Conclusions   **140**

References   142

Species indexes   00

Author index   00

Subject index   00

Preface

During my work on data of the European invasion of the collared dove, it became apparent to me that the existing literature on invasions is either anecdotal or very mathematical. Also, ecologically valuable information exists in population biological disciplines outside ecology proper, such as in epidemiology, population genetics, and biogeography. Clearly, a comprehensive approach might be helpful to students of all these disciplines in providing them with a common denominator of which they might be unaware. Moreover, examples from various disciplines involved can give an idea of work done beyond one's own horizon and thus open up new vistas of understanding.

In this book I attempt to give the multidisciplinary overview required. My approach is to use as many examples as possible, thus showing general principles. Moreover, the processes concerned are quantified to allow mathematical treatment. The mathematics involved is kept to a minimum; only a knowledge of basic statistics is required. Concepts and principles are explained in biological terms and derived from examples given in the text. Results of their application are shown graphically and as maps to facilitate their utilization. It seems to me that this is the only way to bridge the gap between basic biological observations of the processes concerned and their sophisticated mathematical treatment.

I thus hope to bring several fields of research together after their divergence brought about by present-day specialization.

R. Hengeveld
1989

Acknowledgements

Among the many friends who have shown interest in this book, I would like to thank a few in particular. First, the former scientific manager of the Research Institute for Nature Management, Professor R.A. Prins, who gave me the space and much of his encouraging interest to do the work. Mr. R. Wegman made a very good job of drawing the figures. Professors John H.B. Birks and Daniel Simberloff read the entire text and made useful remarks. Last but certainly not least I am grateful to Dr. Francis G. Gilbert and Mr. Frank van den Bosch for their rigorous and most detailed comments. Apart from improving the English, Francis' criticisms often put me in quite a difficult position; I've done my best to meet most of his comments, to avoid mistakes and to make the text easier to understand. Frank was equally helpful and critical during the process of writing, particularly from the mathematical viewpoint; often my understanding of this complicated matter proved not to be up to standard. I've learned much from him, particularly during the simulation of the invasion of the collared dove.

Not only Claire and the boys, but also my parents always showed a keen interest in both the personal and the scientific side of writing; this, I imagine, only few authors can enjoy.

Introduction

Invasions are typically spectacular phenomena, characterized by rapid spread of species over considerable areas and are often regarded as rare and threatening events. Yet, closer study of species ranges shows that a species' internal geographical abundance distribution is usually dynamic rather than static, even when the range margins are stable. This suggests that highly dynamic processes can have a stable, i.e. apparently 'stationary', outcome. Viewed in this way, we can conceive of invasions as temporary disturbances leading to spatial dynamism. Viewed even more broadly, species ranges can be regarded as being in a constant state of flux, both internally and externally. The external aspect is expressed by range shifts, expansions, or contractions. Invasions are then normal, rather than exceptional features of more general range dynamics, their frequency depending on the degree to which a species' properties and requirements match the topography and dynamics of its environment. Moreover, as all species ranges are dynamic to some extent, it is impossible to discriminate between typical invaders and non-invaders among them.

In this book, dynamic concepts are introduced into biogeography and spatial ones into ecology. Also, by using mathematical models from epidemiology and human geography, generalization is feasible, bringing various disciplines together. However complicated the underlying mathematics may be, the basic ideas are simple. We will therefore concentrate on underlying ideas, as well as on practical measures derived from them. However, attempts to integrate various views can never erase their boundaries; rather, they serve to facilitate consultation between the different parties concerned.

The first two chapters contain general information; in Chapter 1, invasions are looked at from the viewpoint of various biological disciplines and Chapter 2 gives some examples. In the next three chapters, various components of invasion processes are formulated mathematically, assumptions made are explained and practical measures derived. Chapter 6 describes the ecological boundary conditions to spatial progression of invasions and explains methods for detecting and estimating the relevant parameters. Chapter 7 gives two examples of simulating invasions, one applied to Neolithic farmers in Europe and the other to stripe rust in a local wheat field. This chapter is technical and can be skipped, as some familiarity is assumed with the original papers. As all the examples given in these chapters are unconnected and pertain to a wide variety of both animal and plant species, Chapter 8 gives the analysis of the invasion of a single species, the collared dove, into Europe during the last half-century. It also puts this invasion in the wider perspective of range dynamics. Following this wider perspective, Chapter 9 analyses the stochastic structure of invasion fronts in great detail taking the spatial and temporal progression of rabies as an example. Finally, in Chapter 10 invasions are interpreted as ecological processes, occurring on a broad spatio-temporal scale. In this chapter a non-equilibrium theory is confronted with current equilibrium theories based on the paradigm of the balance of nature.

In the near future the dynamic nature of species distributions will become more and more apparent, partly because of increasing knowledge about geographical processes both during the Holocene as well as in the past few decades, and partly because of species' responses to the effects of human activity on the environment. Man affects the biosphere in two ways, through the immediate destruction of habitats, and through bringing about long-term changes in the atmosphere. Although most people are aware of present-day habitat destruction, only a few realize the full import of atmospheric changes. Although atmospheric changes are more difficult to detect, recent work indicates that we may expect dramatic changes soon, together with effects on both terrestrial and marine conditions. These altered conditions will, in turn, cause large-scale redistributions of species in geographical space.

Knowledge about the potential spatial dynamics of species will thus become of increasing importance. I hope that this book can help in designing detection, monitoring and warning systems, as well as arouse theoretical interest in species dynamics on broad spatial and temporal scales.

PART ONE
Theory

The first five chapters of Part One give background information on which invasion theory is based. This information is partly empirical and partly mathematical. Because of the divergence of the various disciplines involved, ideas and terminology have diverged as well, sometimes making students of one discipline unaware of similar or identical approaches in others. Chapter 1 therefore puts several disciplines under the same heading again. Chapter 2 does the same, but concentrates on examples that have been worked through, thus introducing several important concepts and processes. These examples show that quantification is necessary and Chapter 3 gives the relevant measures. The next two chapters concentrate on two components of biological invasions; diffusion of dispersal propagules and population growth following settlement.

One
Perspectives of biological invasions

Invasions are known in different areas of population biology, though often under different terms and stressing different aspects. For example, in epidemiology, where spatio-temporal aspects of disease incidence are studied, invasions are known as epidemics, whereas in palynology, invasions are called migrations. Ecologists talk about species introductions or colonizations, or, more locally, about patch dynamics. In population genetics texts it is often easiest to look for waves of advance as these designate the spatial spreading of alleles. The continuous movements of locusts over large parts of North Africa are locally felt as outbreaks but, on the continental scale, they are regarded as swarms drifting about on air currents. In mathematical treatments, invasions are described by diffusion equations and Markov chains.

Yet all these terms pertain to more or less the same process. Before describing the aspects that the various processes share, we first consider them separately to see what is characteristic for each discipline.

1.1 The ecological perspective

Ecologists often distinguish sharply between dispersal and migration, dispersal being a short-distance, non-directional process and migration a long-distance, directional one. However, as it is difficult to decide what are short or long distances, and non-directional and directional movements, much confusion remains. This is because the distances species cover cannot be compared in absolute terms as dispersal capacities are too different. Moreover, short-distance movements can be directional and long-distance ones non-directional. Therefore, I will not distinguish migration from dispersal; if the process is not clear from the context or from the observations themselves, the use of sharply defined terms would not improve matters anyway.

Migration or dispersal is an undervalued process in ecology and communities or populations are, consequently, often considered spatially stable and discrete. In this view, communities would consist of both qualitatively and quantitatively co-adapted species mutually controlling their composition by controlling each other's abundances. Foreign species would have difficulty in penetrating established communities but, if successful, they would disrupt their structure until the established species could control the invader's density and thus establish a new multi-species equilibrium (e.g. Elton, 1958). 'Closed' communities would be less invadable than 'open' ones but, through circular reasoning, predictions on the degree of closedness or openness depend on their invadability! For their part, not all species would be good invaders, some species having properties that enable them to invade foreign communities more easily than others (Leston, 1957). What these properties are is not clear, although many suggestions have been made (e.g. Baker and Stebbins, 1965; Grime, 1986; Safriel and Ritte, 1980, 1983; Simberloff, 1981b).

Species mobility is important in the colonization of islands or new land, such as young polders or volcanoes. Initially, much space remains unoccupied resulting in low competition pressure both among the established species and with recent arrivals. Then an equilibrium is formed between the number of immigrants and that of the established species which eventually may die out due to excessive competition pressure (MacArthur and Wilson, 1967). As competition pressure will relate negatively to the available area, this could explain the positive relationship between the number of species and island size.

This species equilibrium theory has greatly stimulated spatial analysis in ecology and recently resulted in the theory of patch dynamics (Pickett, 1980). In patch dynamics, natural systems are seen as mosaics of local, suitable patches, colonized by various species at different times. Eventually these species die out, thus giving individuals of the same or of other species the possibility of colonization. As various patches differ both in quality and age, highly intricate, spatially dynamic systems can result determining the species composition of the area as a whole.

1.2 The population genetic perspective

As survival probability depends on regional patchiness and on patch dynamics, species either have to adapt to this spatio-temporal heterogeneity or die out. The spatial expansion of new traits concerns another population genetic process, that of centrifugal progression, often operating on broad spatial scales. Fisher (1937) formulated his advancing-wave model to describe this progression in which he combined spatial diffusion with population growth. This combination is essential since, without local population growth occurring, the expansion fades out as local densities become too low. Thus, after reproduction some of the offspring can invade surrounding areas, preventing the wave from exhausting itself.

This theory concentrates on the spatial progression of qualitatively advantageous traits, but does not consider effects of ecological factors and processes apart from exponential growth.

1.3 The epidemiological perspective

As in population genetics, part of epidemiology is also concerned with radial spatial progression, albeit that of plant and animal diseases rather than genes. As such, it is therefore a part of ecology although it has developed independently, since the systems studied are unnatural because of their spatial and qualitative homogeneity. As the spread of diseases resembles the spatial advance of population genetic traits, epidemiology combines aspects of ecology and population genetics. Important ecological aspects are seasonality of the disease and its host and include the dependence of the disease on climatic factors, the degree of habitat fragmentation, and the impact and ecology of possible vectors. As in the advancing-wave model, population growth is relevant in epidemiology; in fact, it is its principal concern (cf. the review by Heesterbeek and Zadoks, 1986).

Epidemiology differs from both ecology and population genetics by emphasizing spatial transfer mechanisms. Thus, when a disease is transmitted by another organism, its vector, it can be eradicated by attacking this vector. Other means of disease transfer depend on the variation in host susceptibility and in the spatial separation of these hosts. Barriers or quarantine measures thus help in controlling the disease through fragmenting its habitat. Epidemiology, being concerned with disease, therefore differs from ecology and population genetics in emphasizing practical control, rather than theory and conservation.

1.4 The biogeographical perspective

Biogeography studies biological patterns and processes occurring on a broad, geographical scale. It concerns a wide range of objects from geographical variation in genetical and physiological traits within species to the distribution of higher-level taxa and differences between continental biotas, irrespective of taxon identity: often, though not necessarily, the processes are broad-scale in time as well.

Because of its comprehensive interest, biogeography comprises the ecological, population genetic and epidemiological aspects mentioned, only differing in the breadth of scale and the objects concerned. Thus biogeographers, like ecologists, are interested in short-distance colonization, but unlike them, also in species introductions on islands and continents. Similarly, they are interested in disease progression after its introduction into other continents or in genetic differences determining resistance levels on different continents, as in Dutch elm disease.

Thus biogeography stands apart, because of its comprehensiveness rather than because of constraints such as those found in the previously mentioned disciplines. Although invasions as geographical processes are often considered exclusively of biogeographical interest, various aspects are also studied in other disciplines.

1.5 The mathematical perspective

Just as with biogeography, the mathematical approach to invasions is comprehensive; it differs from biogeography in its level of abstraction and generality. It defines invasions in terms of processes which are also found outside population biology, such as diffusion in chemical systems or population growth in human demography and geography and it is not concerned with taxon identity or with the history of areas, continents or oceans. But because of the differences between the inanimate diffusion systems and biological ones characterized by, for example, growth, additional parameters should be included in the models. Similarly, various diffusion models can be formulated when the diffusion away from some liberation point follows a general, proportional decay or when the individual's behaviour affecting dispersal determines spatial spread.

Thus, mathematical models deal with general processes but become biologically more specific when more biological parameters are included. Yet every parameter added reduces the model's generality. Describing invasions using mathematical models should thus be done with caution; too great a biological specificity makes the results anecdotal whilst too great a mathematical generality makes the models inapplicable.

1.6 Terminology

Because of the variety of disciplines involved in the study of broad-scale spatial spread, a plethora of terms have come into use. Sometimes, different terms are utilized for the same process, whereas in other instances one term can have different meanings depending on the context or on the author using it.

Usually, invasions are said to occur when a species enters a region where it was absent before. Sometimes the process is restricted to entering other continents, rather than other parts within the same continent (e.g. Elton, 1958). At the other extreme ornithologists use the term invasion synonymously with irruption for irregular, temporal occurrences of birds outside their normal range without permanent settlement there (e.g. Drost, 1962). Finally, we can use the terms to indicate the advance of population waves of high density into areas with low density, in a sense similar to military invasions, though in a different way. Many differences in definition depend on the scale of investigation; some definitions are broad and some are fine, some are rigid, accepting only presence-absence data, whereas others are relaxed, describing quantitative shifts. Often no sharp, principle distinction is made between dispersal, migration, irruption, colonization, epidemic, or range expansion, nor between invasions after a first introduction on a continent, and recurrent recolonizations of regions or fields, even after sometimes rather short time intervals. All existing concepts and their definitions then merge into each other.

Here I would like to distinguish between spread as the spatial process of moving individuals or propagules and spread plus settlement of these individuals or propagules and subsequent growth of the resulting population. In this sense irruptions of birds should not be confused with invasions, as they characteristically concern a temporary spread without subsequent settlement and population growth. In those cases where settlement and growth follow spatial spread we can use the term invasion for the whole process. Similarly, dispersal, migration and nomadism concern the aspect of spatial transfer of individuals or propagules only, just as the process of diffusion discussed in Chapter 6. Contrary to nomadism, dispersal and migration involve spread which does not follow changing ecological conditions in space, dispersal concerning unintentional spread and migration intentional. The terms epidemic, colonization, invasion and range expansion include both the spatial process and those of settlement and population growth. Epidemics always concern diseases, whereas the other terms do not. Colonization is often used for relatively fine-scale processes and invasions and range expansions broad-scale ones. Although invasions are often, though not exclusively, used when the species has a negative impact on the species that were already present in the area, I will not make this distinction here.

Finally, some terms are used to describe the result of the spreading processes, such as dispersion, geographical distribution, range and range extension. Of these, dispersion concerns fine-scale distribution patterns and the others broad-scale ones. Geographical distribution is rather a loose term including several parameters of a species' geographical occurrence, such as the location and breadth of occurrence. Range is more strictly defined as the total spatial span of a species' occurrence, which can be defined more closely using adjectives, such as range location, range size etc. Range extension can preferably be used for the added part of a range after range expansion, although it sometimes refers to the total span or to the expansion as well.

1.7 Conclusions

Invasions can be studied from widely different viewpoints and with different aims, as well as at different spatio-temporal scales and levels of abstraction and generality. But identifying these differences should not keep the various disciplines apart; their identification should facilitate the integration of these disciplines. To complete a puzzle, we should know its pieces.

Concentrating, for example, on ecological processes would leave the aspect of broad-scale spatial progression, studied in population genetics, epidemiology and biogeography, out of consideration, and studying only the geographical wave progression would leave fine-scale ecological species maintenance poorly understood. Also, for species conservation it may be necessary to enhance the spatial dynamism of one species and to lessen that of another, requiring both the ecological and the epidemiological approach.

To improve both theoretical insight and model applicability, it is necessary to integrate information and understanding available in various disciplines. The next chapter takes examples from each, but does not deal with their mathematical modelling. This forms the subject of the rest of the book.

Two
Examples of invasions

This chapter gives examples of invasions from temporally broad-scale to fine-scale ones taken from palynology, population genetics, ecological biogeography and epidemiology. The broad-scale invasions are the recolonization of Europe and North America by tree species after the ice ages, and the fine-scale ones spatial disease progression. Temporally intermediate processes are illustrated by waves of genetical advance and the spatial progression of a species after natural colonization or artificial introduction. Thus, all examples concern spatial processes and differ only in temporal scale.

2.1 Holocene tree invasions

For a long time, palynology was mainly concerned with describing stratigraphical sequences of local cores; only recently has information from many localities been mapped. As these maps differentiate the distribution of individual taxa for particular time horizons, the spatial advance of the taxa can now be described. Moreover, rates of spatial advance can be estimated, together with the taxon's refugium during the last glacial interval or its direction of migration. These parameters can be compared for the various taxa and within each taxon for the different time horizons (e.g. Davis, 1976, 1981; Huntley and Birks, 1983). Although maps of marine taxa such as foraminiferans have also been constructed, these give too little detail for estimating the same parameters.

Fig. 2.1 shows directions in which tree taxa recolonized eastern North America after the last glacial retreat. *Picea* spp. migrated northwards, *Tsuga canadensis* mainly westwards, and *Castanea dentata* northeastwards. *Fagus grandifolia* has a more complex migration pattern, composed of several consecutive directions. Fig. 2.2 shows the colonization of Europe by *Abies*, having two main phases of expansion at 7500-6000 BP and 5500-4000 BP, one of coalescence around 7000 BP, and a final one of decline accompanied by range fragmentation. Fig. 2.3 shows the number of refugia as a function of

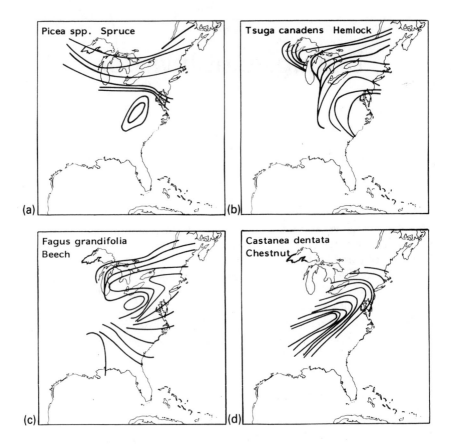

Fig. 2.1 Post-glacial range expansion of *Picea* spp. (a), *Tsuga canadensis* (b), *Fagus grandifolia* (c), and *Castanea dendata* (d). The figures indicate the time of arrival in thousands of years BP (after Davis, 1981).

the number of taxa in each pollen taxon. Their distribution over the refugia in southern Europe was uneven which later added to the complexity of their migration routes. Table 2.1 gives estimated migration rates, showing their heterogeneity.

Thus the taxa migrated individualistically, each having its own refugium, its own migration direction and its own migration rate. In the individual taxa direction and migration rate also varied in time, showing phases of expansion, including coalescence or range consolidation and of decline and fragmentation. Moreover, in some taxa migration has not stopped; witness the still continuing progression of *Picea* into Scandinavia since 2500 BP.

This individualistic geographical behaviour implies that vegetation types did not follow glacial retreat as entities, but that their composition was ephemeral. Moreover, Davis (1986) argued that for the last two million years

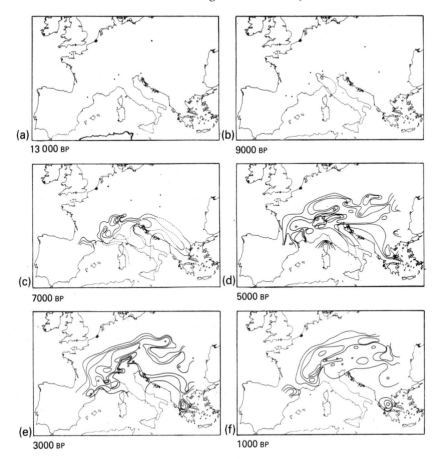

Fig. 2.2 Post-glacial range expansion of *Abies* spp. into Europe. Isopleths (isopolls) connect points of similar relative representation in pollen samples (after Huntley and Birks, 1983).

of the Pleistocene equilibrium conditions cannot be defined for periods shorter than 100 000 years, shorter periods always showing trends. Under such ecologicaly unstable conditions selection can only act against the formation of species-specific co-adaptations (Davis, 1976). Invasions can be more hampered by general effects of established species on their environment than by the structure of the community itself.

2.2 A genetical wave of advance in man

Genetical waves of advance concern the spatial progression of genes within geographically established species. The advantage new genes have over established ones depends either on changes in ecological conditions or on

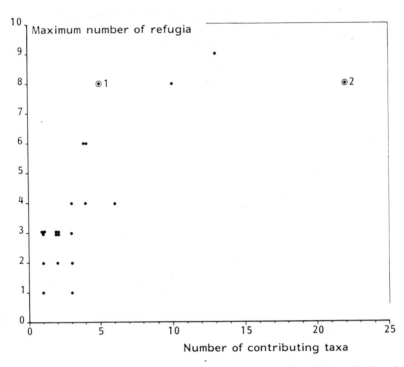

Fig. 2.3 Number of taxa found in various numbers of refugia during the last glacial. The numbers of taxa within *Ulmus* (1) and *Quercus* (2) may be too low or too high, respectively, either due to taxonomic uncertainty or to deviant intraspecific variation (after Huntley and Birks, 1983).

changes in the species' properties themselves. Ammerman and Cavalli-Sforza (1984), for example, argue that the agricultural revolution that marked the end of the Mesolithic was responsible for the change in the genetic make-up of the early Europeans through changes in digestion necessary because of dietary changes (cf. Renfrew, 1987 for linguistic implications). This wave of cultural changes, accompanied by genetic ones, advanced at about 25 km per generation of 25 years each (Fig. 2.4). As a final stage of their analysis, they simulated the process, using frequency distributions of migration distances in a geographical diffusion process, Lotka-Volterra equations for interactions between Mesolithic hunters and Neolithic farmers and local logistic population growth (cf. Chapter 7).

The rate of spread can be measured from maps in two ways; archaeologically using artefacts (Fig. 2.5), and genetically, by mapping the genes for 38 blood groups. The broad similarity between the maps obtained implies that the genetic measure describes the spatial spread reasonably well. Maps were also constructed using ethnographic information in a simulation model

Table 2.1 Migration rates and numbers of certain and possible glacial refugia

Taxon*	Range of observed migration rates ($m\ yr^{-1}$)
Abies (6)	40 – 50 – 300
Acer (13)	500 – 1000
Alnus (4)	500 – 2000
Carpinus betulus (1)	50 – 100 – 200 – 300 – 700 – 1000
Castanea sativa (1)	200 – 300
Corylus-type (4)	1500
Fagus (2)	200 – 250 – 300
Fraxinus excelsior-type (3)	200 – 500
Fraxinus ornus (1)	25 – 100 – 200
Juglans (1)	400
Larix (2)	—
Olea europaea (1)	—
Ostrya-type (2)	—
Phillyrea (2)	—
Picea (2)	80 – 240 – 500
Pinus (10)	1500
Pinus (Haploxylon) (3)	—
Pistacia (3)	200 – 300
Quercus (Deciduous) (22)	75 – 150 – 200 – 500
Quercus (Evergreen) (3)	—
Tilia (4)	50 – 130 – 300 – 500
Ulmus (5)	100 – 200 – 500 – 1000

*Numbers in brackets are the number of native plant taxa contributing to the pollen taxon.

of spatial diffusion for 10 genes, each with two alleles, combined with logistic growth and population interactions with steps of 40, 80, 120, and 160 generations. Migration occurred up to the end of this period because reproduction within the farming population was greater than that of the former hunters; after this Neolithic transition, no comparable migration occurred causing the the pattern of gene frequencies to consolidate. The similarity of the three patterns using archaeological, genetical and simulation experiments using ethnographic information suggests that the diffusion model used is appropriate.

2.3 The red deer and the thar in New Zealand

From 1851–1922, and using 5 strains, the red deer (*Cervus elaphus*) has been liberated 32 times and at 20 places in the northern part of South Island, New

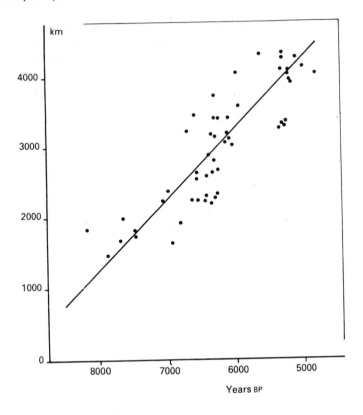

Fig. 2.4 Time of arrival of early farming in Europe from Jericho as the diffusion centre. The drawn line represents the first principal axis (after Ammermann and Cavalli-Sforza, 1971).

Zealand (Clarke, 1971). Only one of these introductions has succeeded, the other 31 liberations all failed. The successful one, a herd of one stag and two hinds, was made in 1861 near Nelson City, but the herd did not spread extensively during the next 80 years (Fig. 2.6(a)). After 1940, however, it expanded at a high rate in all directions but mainly to the south (Fig. 2.6(f)). Fig. 2.7 shows its sigmoid expansion rate with, surprisingly, a maximum of 637 square miles per year between 1910 and 1920 after its consolidation before 1900. Thus, during this decade expansion rate was high, despite a still limited areal extension. Moreover, it slowed down before reaching the limits set by island size. Population size may also have increased dramatically during this period. After 1920 parts of the population started to coalesce, hampering judgements about the species' rate of increase.

During the expansion phase, an inner high-density core containing the bulk of the breeding population could be distinguished from a peripheral

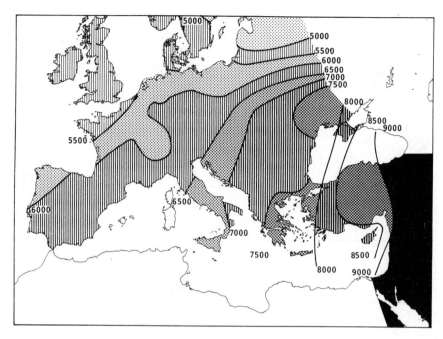

Fig. 2.5 Isochrons of arrival times of early farming based on 106 sites across Europe (after Ammerman and Cavalli-Sforza, 1984).

low-density zone, mainly containing stags and scattered hinds and their calves. Usually, the stags colonized the island 1–3 years prior to the hinds, a difference that ceased in different regions in different years. After this they were integrated in the general range, though still living separately after the rutting season. During the period of rapid areal expansion density of the breeding population declined after which it increased again, the expansion apparently taking many animals from the breeding centres.

The population did not expand radially from the liberation point, but was restricted by both topography and vegetation. The herds typically followed river valleys, colonizing higher altitudes only after regional establishment. Roads and railway lines also formed barriers. Apart from this, areal expansion was restricted after 1920 by the surrounding ocean which may explain the slowing down of the rate of areal increase, the remaining increase being due to filling in of the already extensively occupied area.

A bovid, the Himalayan thar (*Hemitragus jemlahicus*), was also introduced into the South Island of New Zealand (Caughley, 1970). From about the turn of the century onwards to 1919, thar were liberated in various years and at different locations. The number of animals introduced was 29, derived from a gene pool contributed by 13. Their distributions in the high sub-alpine

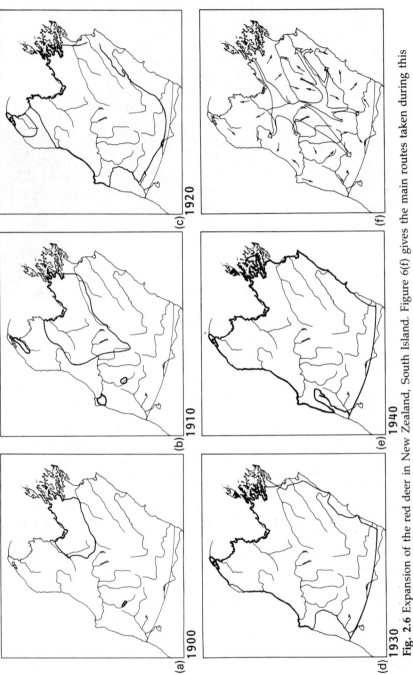

(a) 1900   (b) 1910   (c) 1920

(d) 1930   (e) 1940   (f)

Fig. 2.6 Expansion of the red deer in New Zealand, South Island. Figure 6(f) gives the main routes taken during this expansion (after Clarke, 1971).

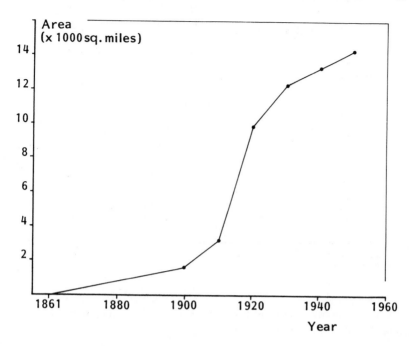

Fig. 2.7 Expansion rate of red deer in South Island, New Zealand (after Clarke, 1971).

zone were estimated in 1936, 1946, 1956, and 1966; after 1966 they were poisoned and shot as control measures which prevented further estimations of range extension to be made.

As in the red deer, male thar were the first to colonize areas surrounding the breeding range, up to 60 miles from the range margin. This early colonization relative to that of the females and juveniles happens prior to the mating season and seems to be the consequence of searching movements made by the males. When the females enter this zone, fecundity scarcely alters due to the Allee effect (cf. Klomp *et al.*, 1964), i.e. only by possibly sparse contacts with males.

Fig. 2.8 shows the overall expansion rate in terms of the square root of area divided by π, i.e. the radial equivalence of the area occupied over time. Taking the radial range expansion, Caughley (1970) applied an equation developed by Skellam (1951), comprising two invasion components, diffusion and logistic population growth. As the curve is straight after having stabilized, Caughley concluded that Skellam's (1951) model applies rather than ones based on exponential population growth. From this, in turn, he concluded that range expansion in thar results not so much from population pressure as from an innate tendency to disperse (cf. also Chapter 8).

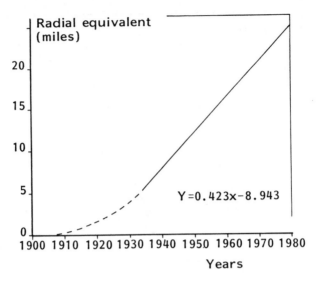

Fig. 2.8 Expansion rate of the Himalayan thar in the eastern population in New Zealand expressed in terms of the increase of the radial equivalent over time (after Caughley, 1970).

2.4 Cholera in North America and measles in Iceland

Cliff *et al.* (1981) distinguished several forms of diffusion, the most important of which are neighbourhood diffusion and hierarchical diffusion. Neighbourhood diffusion results when the disease progresses from an initial source — a patient — to its immediate neighbour — a susceptible (Fig. 2.9(a)). Long-distance diffusion, in contrast, progresses with great leaps, thus bypassing neighbours. In practice, both processes will occur simultaneously, diffusion progressing stepwise, thus forming new infection centres from where it infects neighbouring susceptibles, thus resulting in hierarchical diffusion (Fig. 2.9(b)). In human epidemics, shifts may occur from the first type to the second depending on the increasingly greater ease with which distances are covered.

Such a shift can be seen in three subsequent North American cholera epidemics occurring in 1832, 1848–49 and 1866 and spreading from New Orleans and New York (Pyle, 1969). In 1832 the spread followed main corridors with distance being the main parameter (Fig. 2.10(a)). During the second outbreak of 1848–49 the spread from New Orleans shows both neighbourhood and hierarchical spread components indicating that both distance and urban size determine the process (Fig. 2.10(b)). During the final outbreak of 1866 distance was subordinate to urban size (Fig. 2.10(c)). This shift from distance to urban size as the dominating factor suggests the

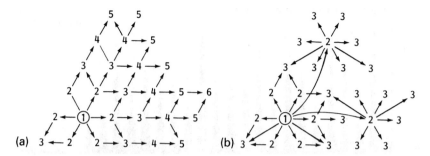

Fig. 2.9 Schematic representation of neighbourhood diffusion (a) and hierarchical diffusion (b). In hierarchical diffusion big jumps are made after which the species spreads with small ones according to neighbourhood diffusion (after Cliff *et al.*, 1981).

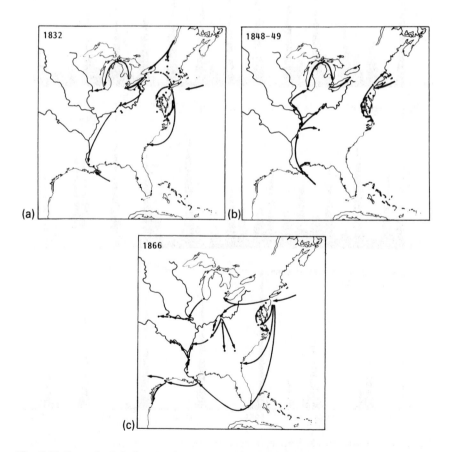

Fig. 2.10 Spread of cholera in the eastern United States in three periods during the 19th century. The figures refer to the day of first reporting (after Cliff *et al.*, 1981).

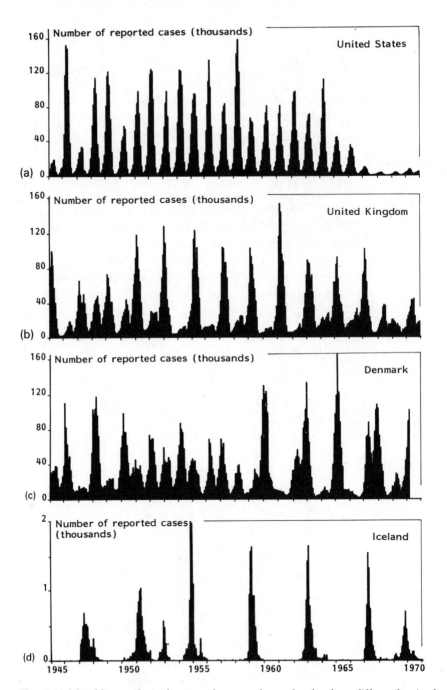

Fig. 2.11 Monthly number of reported cases of measles for four differently sized countries. In all four countries measles has a clear cyclicity, although in Iceland it is non-endemic (after Cliff *et al.*, 1981).

influence on wave velocity of population increase, of lessening travelling difficulties or both.

Another infectious disease, measles, is transmitted directly from a patient to a susceptible resulting in the epidemic fading out after the pool of susceptibles is exhausted. In the meantime, the disease must reach other populations of susceptibles in order to sustain itself. If it can be maintained in a region the disease is endemic to it; if not, it has to be reintroduced after each extinction. Thus, the size of the pool of susceptibles and the transmission probability determine whether a disease occurs endemically or not.

Fig. 2.11 shows that up to the late 1960s in the smallest population (Iceland) measles had to be introduced from abroad, after which it became endemic. In the largest population, that of the United States, it was endemic for the whole period declining only after the start of vaccination programmes in 1965. Here the disease also had the most regular cycle period, cyclicity thus increasing with population size. Fig. 2.12 shows the rate at which measles fades out as a function of island population size.

In Iceland, measles epidemics often started at Reykjavık, spreading from there to other towns or villages which became secondary centres of spread. Thus, measles progressed hierarchically from the largest town down to the smallest settlements and families. The coastal route was an alternative way

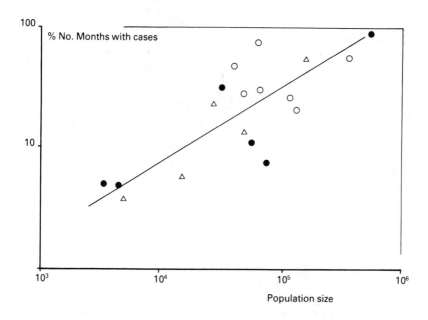

Fig. 2.12 Proportion of months with measles incidences in island populations of different size (after Cliff *et al.*, 1981).

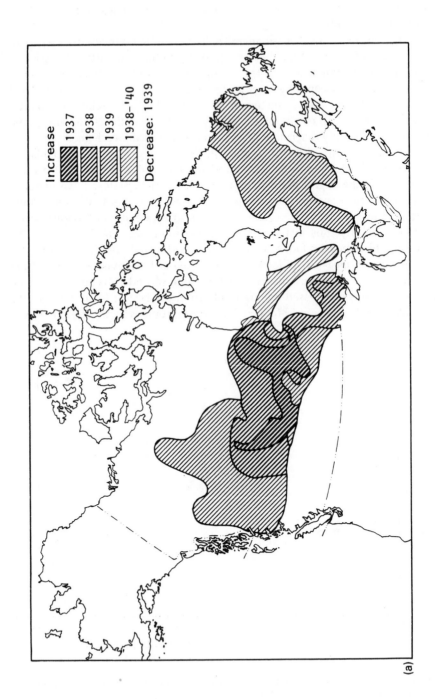

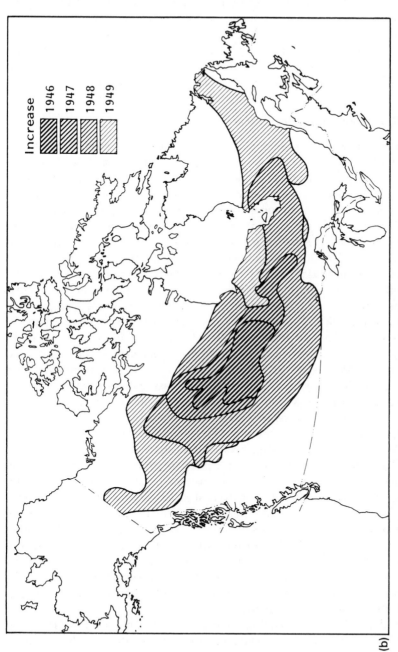

(b)

Fig. 2.13 Expanding wave of population increase of the Canadian lynx following irruption in a central area. This wave usually spreads in all directions, though at different and often irregular rates (after Butler, 1953).

of disease transmission relative to direct over-land routes. During the relatively cold 1920s and 1960s both routes were blocked for part of the year, being one of several socio-economic factors affecting disease incidence.

As in cholera in North America, measles epidemics can be described in terms of diffusion, i.e. partly neighbourhood diffusion and partly hierarchical diffusion. Mainly due to changes in urban and settlement size and to increasing mobility, shifts in disease incidence have occurred.

2.5 Recurrent population waves

In the recurrent waves of high population density of, for example, the snowshoe hare, *Lepus americanus*, across Canada, we can also speak of invasions, although this would probably inflate the term too much. The process of wave propagation, however, may be the same or quite similar: it concerns quantitative rather than qualitative spread. Butler (1953) divided the Canadian fur-collecting area into 63 more or less homogeneous parts and estimated peak years for each sub-area. As for several other fur-bearing species, Butler identified cyclic fluctuations in the snowshoe hare with lows in 1927, 1936 and 1945. Next, he plotted the peak years for each sub-area on annual maps and connected these years by isophasal lines to delineate areas of population increase from those of population decrease or constancy. It then appeared that this hare increases after each low in certain parts of Canada, this phase of increasing abundance spreading out in wave-like fashion in various directions from the area of initial increase (Fig. 2.13(a) and (b)). Butler (1953) suggested that migration from densely populated areas into surrounding, less populated areas is the easiest explanation.

Similar to the snowshoe hare, the cycle of the Canadian lynx, *Lynx canadensis*, starts in a central area from where it spreads centrifugally (Butler, 1953; Elton and Nicholson, 1942). Smith and Davis (1981) analysed fur data of this species using bivariate spectral time series analysis and found that the lynx fluctuates with a 10-year period. Using cross spectral analysis the relationship between the fluctuation pattern of six geographical areas was described next. The analysis showed that, similar to earlier findings, the wave of population change starts in the west central parts of Canada from where it spreads in all directions. Thus, regional fluctuation patterns are synchronized relative to this central or nodal area of origin of the wave. Moreover, after dividing the total time series in two, the nodal area of the nineteenth century, extreme northern Alberta, had shifted 500–600 miles to the southeast to northeastern Saskatchewan or western Manitoba in the twentieth century.

Smith and Davis (1981) did not find a causal explanation of the spreading wave, nor of the shift in its nodal area. Nevertheless, they felt that at least a partially external set of factors whose nature changes over time would be operating. Whatever the cause that eventually is found, the phenomenon of spreading-out waves of advance seems clear enough. These waves are recurrent with particular lengths and their way of advancing can be similar to that of invasions in a stricter sense.

2.6 Conclusions

These examples contain several elements often present in invasions. Holocene invasions of trees show that (1) taxa differed from each other in refugium occupancy, (2) migration rates and directions varied both within and between taxa, and (3) separate parts of ranges coalesce because of, and consolidate after, range expansion, and fragment when they contract. Individualistic behaviour rather than coherence or structural inertia of communities characterizes recolonization. The population genetic progression distinguishes between diffusion, growth and hunter-farmer interactions. A change in socio-economic population structure appears to be the driving force of the genetic wave. The invasion of red deer shows that only one out of 32 introductions of five strains liberated at several locations was successful. The present population of many thousands of individuals descended from three founding individuals only. The demography has three important components: (1) the expanding range consists of a peripheral, low-density zone surrounding a high-density core with most of the breeding population, (2) numerical exhaustion of the breeding population because of spatial dilution, and (3) initial establishment was followed by a period of rapid range expansion. This rapid expansion is followed by population coalescence and by a slowing-down of population growth due to areal limitation. Topography and vegetation constrained expansion direction. The cholera and measles epidemics show historical shifts in the importance of distance and urban size due to both climatic and socio-economic developments relative to the exhaustion rate of the disease. Finally, invasions or invasion-like phenomena can be recurrent with particular cycle lengths. Also, processes recognized in invasions can be similar to those in other processes occurring at other spatio-temporal scales studied independently in different disciplines and often known under different terms.

Three
Measures of areal expansion

As invasions advance from their starting point or epidemic diseases from their inoculi, they spread at radially symmetric rates. However, barriers can constrain this spread in certain directions resulting in ranges of irregular shape. At the fringes of oceans, mountain ranges or deserts, expansion can even become unilateral (Fig. 3.1) and zonally varying climates can cause bilateral expansion (Fig. 2.13).

In all instances, however, expansion rates should be measured in a comparable way and, if necessary, be corrected for any constraints. Three related measures are used, radial increase, areal circumference and the square root of the surface occupied.

3.1 Relative and absolute radial increase

The simplest measure of range is its radius. Radius at known times gives a measure of expansion rate. This rate is expressed by the slope of the regression line (Fig. 2.4). Although this absolute measure is straightforward, Van der Plank (1963) used the relative radial increase, in which the radial length added during a certain time unit of, say, a week or a year, is expressed as a percentage of the total previous radial length. A mathematically related variant is the relative number of spatial units occupied by the invader or infected by the disease.

These relative measures have two related drawbacks, the artificially introduced time-dependence of the results and the increasing relative weight of the previous expansion. Of course the rate of increase during a particular period can depend on the state of, or increase rates in previous periods, but usually this dependence decreases for periods farther back in time. Thus, in the case of time-dependence it should be weighted for the various time periods. Moreover, relative increase lessens for the same absolute areal increase because of the increasing weight of the expanding area after each time interval. Thus, pseudo-asymptotic expansion rates can obscure real slackening rates.

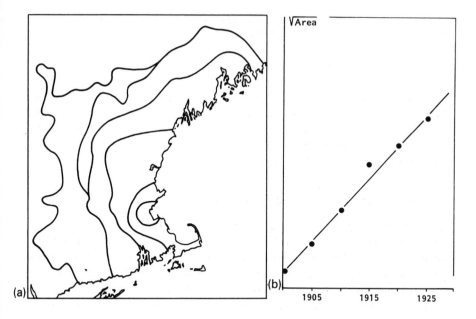

Fig. 3.1 Geographical expansion and expansion rate of *Lymantria dispar* in the eastern United States (partly after Elton, 1958).

Absolute measures avoid difficulties inherent to relative increase rates, although in practice other difficulties, that are also inherent to relative measures, arise when species expand asymmetrically. The red deer, for example, expanded mainly southwestwards and as secondary expansion waves, laterally (Fig. 2.6(f)). Unilateral or bilateral expansions (Figs 2.13 and 3.1) are other examples. The problem of asymmetric expansion cannot be solved by measuring along the principal invasion axis, as this ignores the fanning-out along this axis. More integral measures are therefore needed.

3.2 Areal circumference

Nowak (1975) proposed areal circumference as a more integral expansion measure. Thus, he expressed expansion in terms of areal circumference minus the increase of the invasion front. This rate can be expressed either in absolute or in relative terms as the percentage of 100 km front length, for example. This relative measure, however, approaches a constant value with increasing radius length. Another drawback Nowak (1975) mentions is that range limits are usually unknown at yearly intervals. But when increase rates would be expressed by the regression of areal circumference against time missing values need not seriously affect the results.

A more serious difficulty is the interpretation of circumference as derived from surface area. Using the equation

$$a_r = \pi(r + nk)^2 - \pi[r + (n-1)\,k]^2,$$

assumes circular increase, whereas areal circumference increases more the larger the deviation from circularity. As longer limits require more individuals for a given areal increase than shorter ones, they will progress more slowly at similar reproduction rates. Surface area is less likely to be mis-interpreted and is the preferred measure here.

3.3 The square root of the area occupied

The size of a population's range is often proportional to its area and as it expands its area will relate to time in a simple way. In fact, the square root of area increases linearly with time (Skellam, 1951). Thus, the radially more or less symmetric range expansion of the muskrat in Europe is constant over almost the whole of the period concerned (see Chapter 5 for the explanation of some deviating years), which also holds for unilateral invasions (Fig. 2.13).

However, this assumes that the increase rate is constant, which does not always apply. Figs. 3.2 and 3.3 show that the Japanese beetle *Popillia japonica* expanded curvilinearly before 1925 and linearly after that time. Presumably, following its introduction, the species adapted to its new environment after which it started to expand. The kind of processes involved during this initial

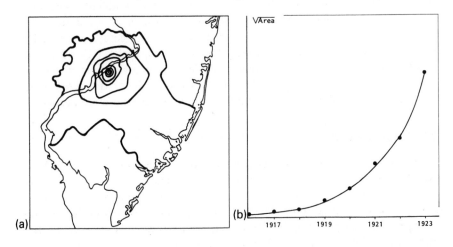

Fig. 3.2 Geographical expansion and expansion rate of the Japanese beetle *Popillia japonica* in the eastern United States. Its initial expansion rate was curvilinear in contrast with later stages when it became rectilinear (partly after Elton, 1958).

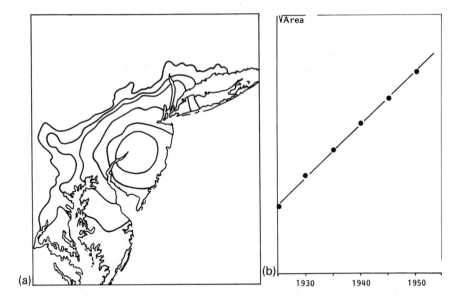

Fig. 3.3 Geographical expansion and expansion rate of the Japanese beetle *Popillia japonica* in the eastern United States at a broader spatio-temporal scale than in Fig. 3.2 (partly after Elton, 1958).

period are unknown but synecological or genetic adjustments may be involved (see also Chapter 5).

Approaching habitat saturation of the newly colonized area, population increase may slow down again, asymptotically reaching a certain density level for the prevailing conditions. The mechanism often thought to be involved is intra- or interspecific competition, the intensity of which increases with increasing population size. Thus, the decreasing expansion rate of the red deer in New Zealand (Fig. 2.7) can result from increasing densities because of the impossibility of expansion beyond the fringes of the island; had the island been larger, the species could have expanded at the same rate for a longer time. Significantly, the curve starts to level off in 1920 when the fringes of the island were reached and the population started to coalesce.

3.4 Conclusions

Of the various measures discussed, the one using the square root of area as a function of time is the most practical and the easiest to apply and interpret. Moreover, its equation has been derived from those of processes based on population growth and on chance distributions resulting from different modes of dispersal. These population processes, growth and dispersal, are discussed in the next two chapters.

Four
Population growth

Different from, for example, diffusing molecules, biological propagules can reproduce after their arrival in new areas. Thus, one component of invasions concerns the spatial transmission of propagules and a second is growth of local populations through reproduction. Reproduction of propagules is important as it prevents the invasion rates from dropping gradually to zero by mere shortage and eventual lack of propagules. On the contrary invasion rate may, to a large extent, be determined by the reproduction rates of the various local populations.

However, depending on ecological conditions, reproduction rates vary within species. They also vary among species because of their specific properties, such as seed number, or the determination of reproductive output by the species' physiology itself. These properties depend partly on ecological conditions and are partly under genetic control. Thus, ecological and genetical conditions, together with their interaction, determine a population's growth rate. Moreover, they determine the type of growth, formulated in various growth models. Of these, I discuss two types, exponential growth and logistic growth.

4.1 Exponential growth

Population growth can be non-reproductive resulting from immigration from external sources and occurs in, for example, monocyclic crop diseases with only one generation per growing season. This growth type is called linear growth (Zadoks and Schein, 1979). In polycyclic diseases, by contrast, having more than one generation per growing season, the new generation starts reproducing parallel with the parent generation. Growth is then exponential as long as the environment is not limiting. Thus, overlap of generations within the sampling period distinguishes these two types.

Thus, in the first generation there are N_0 reproducing individuals, in the second $N_0\lambda$, in the third $N_0 \times \lambda \times \lambda = N_0 \times \lambda^2$, etc. Over t generations there are $N_0\lambda^t$, or N_0e^{rt} individuals. Here, $\lambda = e^r$ where e is the base of

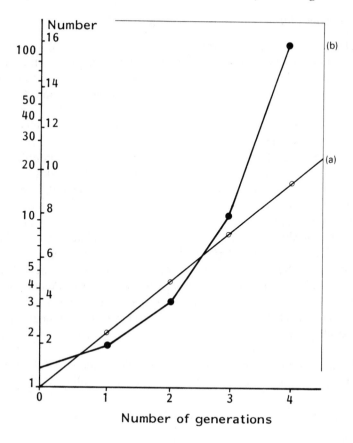

Fig. 4.1 Population growth of bacteria plotted linearly (a) and semi-logarithmically (b).

natural logarithms of reproduction rate *r*; the exponent of the population's growth rate being the difference between birth and death rate, assuming no migration. Plotted as the number of individuals as a function of the number of generations, the line describing the numerical increase is curvilinear which becomes straight when plotted logarithmically (Fig. 4.1). This regression line expresses the logarithmic transformation of the above formula, $\ln N_t = rt + \ln N_o$.

Populations soon increase extremely fast when they grow exponentially. Thus, starting with a population of 10 individuals having 10 reproductive individuals as offspring results in $10 \times 10 = 100$ individuals in the first generation, in $10 \times 10^2 = 1000$ in the second, and 10^6 in the fifth. As this does not happen in nature, we may infer that the basic assumption of unlimited growth is unjustified. Sooner or later the environment limits growth in some way.

4.2 Microepidemics

Populations cannot grow indefinitely: at some point the environment stops them in one way or another. Two things can then happen; either the population dies out because it has exhausted its resources, or it stops growing after reaching a certain population level. The present section discusses the first process and the remainder of this chapter the second.

After settling, populations can exhaust their local environment causing them to die out. In the meantime they must have formed propagules to reach other sites where they can settle and build up new populations from where, in the next cycle, they spread again. On a fine scale, therefore, the process is highly dynamic consisting of cycles of settlement, local extinction and dispersal (see, for example, Carter and Prince, 1981). On a broad scale, for example that of the species as a whole, the process can be static, depending on the number of sites, dispersal capacity, etc. Thus, Fig. 1.12 shows the relationship between the decline of measles and island population size, the largest islands having the smallest chance of this happening. Also, in the largest population, that of the United States, measles is endemic and has the most regular cycles, whereas in the smallest population, that of Iceland, there is no cyclicity. There the disease is epidemic and is introduced occasionally from abroad after which it exhausts its pool of susceptibles and dies out again (Fig. 1.11).

Measles is endemic in the Brazilian state Rio Grande do Sul. In this state, weekly occurrences were assembled from 232 municipalities during 1971-1980

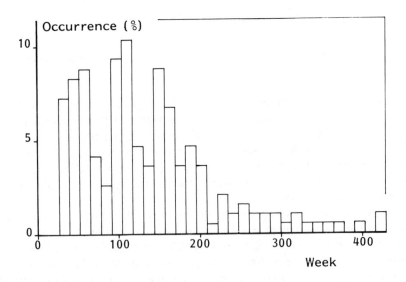

Fig. 4.2 Percentage occurrene of inter-epidemic periods in measles incidence in Brazil, expressing periodic returns of the disease (after Infantosi, 1986).

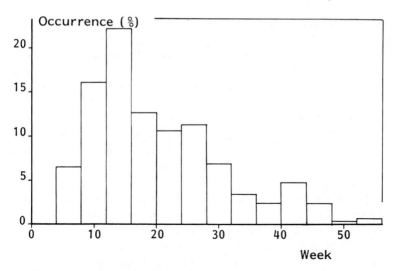

Fig. 4.3 Duration of 346 individual measles outbreaks in Brazilian municipalities (after Infantosi, 1986).

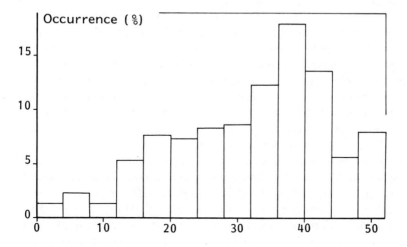

Fig. 4.4 Time of occurrence within the year — seasonality — of measles incidence in Brazil (after Infantosi, 1986).

(Infantosi, 1986). This data set is so large that frequency distributions of periodicity and duration of the microepidemics can be constructed. Fig. 4 2 shows that outbreak periodicity, expressed by the number of inter-epidemic weeks, varies considerably although 86% of the periods are shorter than 4 years. Apart from periodicity, the duration of an epidemic outbreak could also be derived. Figure 4.3 shows that this characteristic also varies extensively, although 50% last 8-20 weeks. Moreover, as in many animal and plant species, measles occurs seasonally within the year. Figure 4.4 shows that 95% of the microepidemics occur between the beginning of autumn and the end of spring. This means that environmental factors external to the disease and host system also play their part.

Thus, considering measles cases for a large state and over several years shows that the process is stochastic with great variation in basic characteristics. The outcome as seen over the whole state and over a long period is that measles is endemic, annually accounting for large numbers of deaths. Although comparable data do not exist in biology, it is certain that the same holds for similar characteristics of the dynamics of animals and plants. We can expect that the variances will often be even larger, depending on dynamic and spatial properties of the species and its habitat.

This is typical for stochastic processes, such as invasions, resulting from new living areas being opened up and random movement of individuals or dispersal propagules transferred between old and new areas. In fact, the process ecology deals with is even more complicated than that of measles propagation because of variation in, for example, habitat (island) quality, inter-island distances or island size. Apart from this the process is usually non-stationary, which means that trends in environmental quality occur both in space and in time and also that the fitness of individuals can vary.

Viewed from this perspective and at a fine spatio-temporal scale, we can expect populations to die out frequently as a consequence of the stochasticity of the process. Extinction is also a population process of interest in addition to population growth and stability. Invasions, for their part, demonstrate a fourth kind of process continuously operating to compensate for local extinctions or for geographical shifts in habitat quality. Depending on the scale of resolution chosen, invasions or epidemics are continuous processes or consist of spatial foci and temporal microepidemics. These microepidemics can be characterized by various statistics, such as their periodicity, duration or their interval time. Moreover, as units of propagation they are subject to environmental variation shown up by their seasonality. All these statistics vary, depending on properties of the species and on those of their environment, making the process stochastic.

4.3 Logistic growth

Another way in which populations can cease growing is by gradually intensifying growth limitation with increasing abundance. The best-known formulation is the logistic growth equation in which the exponent r of exponential growth is replaced by $r-bN$ where b is a scaling constant. Thus, this value decreases for larger values of N, up to a level N_{max} or K, where it equals zero (Fig. 4.5(a)). This level is called the environment's carrying

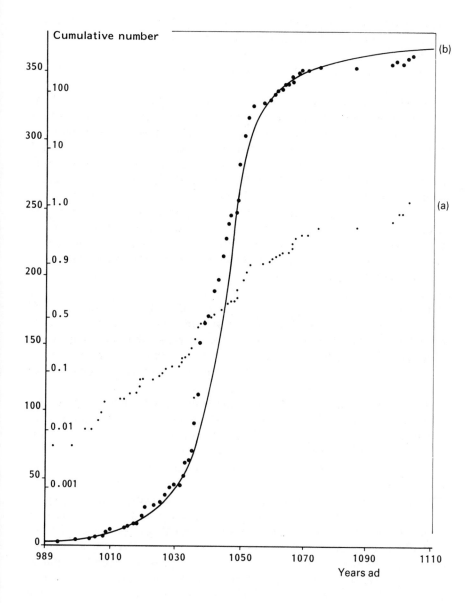

Fig. 4.5 Cumulative distribution of log cutting in an Indian village in New Mexico between AD 989-1112, plotted as logistic probabilities (a) and linearly (b) (after Eighmy, 1979).

capacity beyond which further growth is not possible in the long run. The population's rate of increase can then be formulated as

$$dN/dt = rN(K - N/K)$$

which is known as the logistic equation. As N approaches K, the right-hand side of the equation, and hence the population's growth rate, becomes zero.

There are two ways to describe this relationship, analytical and graphical. The analytical method is complicated and entails iterative estimation of K (cf. Poole, 1974). Graphically, one plots the percentage of individuals cumulated over time on logistic probability paper; when the logistic equation applies, the resulting curve is straight. As Fig. 4.5(b) shows, the seemingly good fit in Fig. 4.5(a) deviates during two periods from a straight line because of slightly too rapid growth. These deviations seem significant because the technique of cumulating percentages averages out experimental error. A drawback of this graphical technique is that the results depend on the length of the time-series chosen.

Although the use of the logistic equation is widespread there are two difficulties with its applicability, concerning its predictability and assumptions. The first is the most serious difficulty; as carrying capacity cannot be estimated, K is an *ad hoc* measure, rather than one that can be obtained — and hence predicted — analytically. It only holds for the observations from which it is derived. Consequently, several mechanisms can lead to the same result, such as competition for food or predation pressure. In fact, sigmoid curves can be obtained from entirely different processes making it impossible to infer logistic growth from such curves without additional information (cf. Pielou, 1974). If it applies, the parameters in the equation should be tested as to their actual operation.

Thus the interpretation of the logistic equation is varied, and even its applicability has been disputed for a long time (Kingsland, 1982, 1985). Although it has the advantage over another growth equation, that of Gompertz, in that it puts an upper limit to growth, it does so through the parameter K for the environment's carrying capacity. Apart from the fact that K cannot be predicted independently, it does not account for density oscillations occurring after the carrying capacity has been reached. But as carrying capacity depends on the integral species response to the joint action of all biotic and abiotic factors, K is not unique for a species, but varies with local species composition and environmental conditions. Thus, Gause (1932) found that K varies according to an optimum curve as a response to temperature, which would explain similar responses along local or geographical gradients (Whittaker, 1967). Its geographical variation is expressed by the two-dimensional optimum surface of intensities over the species range (Brown, 1984; Hengeveld and Haeck, 1981, 1982).

Another, related, interpretation of K is that it represents the maximum value of the range of growth increments at any period of time. When these increments are not cumulated but plotted straightaway, that is when growth rate is plotted against time, another optimum curve results which is the first derivative of the logistic equation. The skewness of this curve indicates the asymmetry of the sigmoid. Varying K according to an optimum distribution relative to an environmental factor means that both the mean and the variance of the distribution of the first derivative vary accordingly.

The logistic growth equation makes six assumptions, one or more of which do not apply under field conditions:

1. the population has a stable age distribution
2. the rate of population increase per individual decreases linearly with increasing population size
3. the rate of population increase per individual decreases instantaneously with increasing population size
4. the environment is constant
5. because of its uniformity, the environment affects all individuals at all stages equally
6. the environment does not affect the equality of mating probabilities among individuals.

But environments are often not constant, affecting individuals of different ages differently and hence the stability of the age distribution. Also, demographic responses are usually not instantaneous, but lag behind changes. Apart from these and other difficulties carrying capacity itself is not constant but varies, assuming indeed that it is reached at all. Crops, for example, are often harvested before a disease can reach its carrying capacity. This holds similarly in many seasonally varying species under natural conditions and for species invading a patch, thus preventing other, already established species from expanding.

Thus, for one reason or another, logistic growth does not occur under many field conditions when sigmoid growth curves can be understood in terms of truncated, para-logistic growth (Zadoks and Kampmeier, 1977).

4.4 Logistic growth on different spatial scales

Fig. 4.6(a) shows the spread of the potato blight (*Phytophthora infestans*) from field to field, which is similar to, if distinguishable from, local increase within fields (Fig. 4.6(b)), the only difference being the time elapsed in both cases. For the between-fields expansion, the period was 1 June–13 July 1953 and that within the fields 1 July–1 August. In fact, the last field was infected in the middle of July when the increase in local fields had hardly started, most fields only being lightly infected. But increases in local fields also involve spatial spread, implying that spread did not stop until the beginning

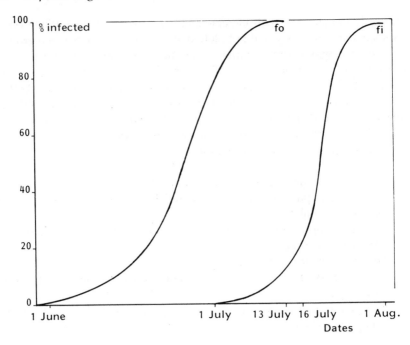

Fig. 4.6 Between-fields (fo) and within-fields (fi) infection rates of Dutch potato fields with *Phytophthora infestans* in 1953 (after Van der Plank, 1963).

of August. Two types of spread can thus be recognized; a brief broad-scale spatial spread over the various fields, and a spatially fine-scale one taking much longer (cf. also Heesterbeek and Zadoks, 1986).

Other observations on potato blight indicate that distinctions can also be made between steep-gradient spread and shallow-gradient spread, the first type within foci immediately around inoculi and the second type on a spatially broader scale between inoculi. But bipartitioning of spreading types is simplistic, with many more spatial scales being possible. For potato blight, Van der Plank (1967) mentions several ways of water and air dispersal, ie. by dew on still nights, by splashing from sprinkler irrigation during calm days, by driving rain in a gale, or by aerial turbulence. Also, human transport can be by movements of a sprayer, or in cargoes from Scotland to South Africa! He concludes that all that is typical of dispersal patterns is their diversity. Moreover, it would be wrong to think in terms of the most efficient dispersal strategy; long-distance dispersal being required to start new colonies and short-distance dispersal to begin new lesions. One should therefore think in terms of combinations of processes operating on different spatio-temporal scales.

Combinations of several dispersal mechanisms operating simultaneously, however, make it difficult to predict particular types of spatial advance insofar as this is influenced by population growth. Usually, logistic growth is assumed for practical reasons rather than for theoretical or practical understanding; the logistic equation is then used as the easiest description of population growth. Then it depends on the importance of short-distance dispersal relative to that of long-distance dispersal and thus how far the wave of advance reflects population growth (Mollison, 1977). If short-distance dispersal prevails the advancing wave adopts characteristics of local population growth; if, however, long-distance dispersal prevails the wave front breaks down, not showing general characteristics.

4.5 An example: the European starling in North America

From the turn of the century, the European starling, *Sturnus vulgaris*, invaded and spread across North America, thus building up a population of 200 billion birds. After several attempts to introduce it in North America, a population was finally established in New York City, starting with 60 individuals released in 1890 followed by another 40 in 1891. During the next decades, it expanded centrifugally (Fig. 4.7). Later its expansion rate differentiated spatially, becoming highest in the northern parts of its new range and lowest in the southern parts (Fig. 4.8). Its areal increase is sigmoid (Figs 4.9 and 4.10). The initial increase in expansion rate can follow from the diffusion process, and the latter partly from truncation due to exclusion of the Canadian part of its range, partly from reaching the Great Plains without much growth and from eventually reaching the deserts in the west.

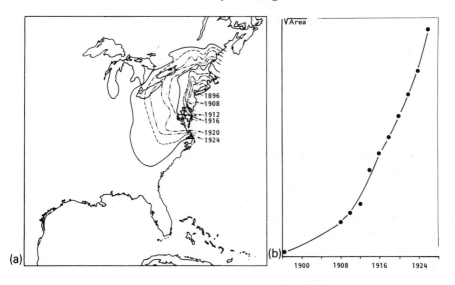

Fig. 4.7 Geographical expansion and expansion rate of the European starling in eastern North America (partly after Elton, 1958).

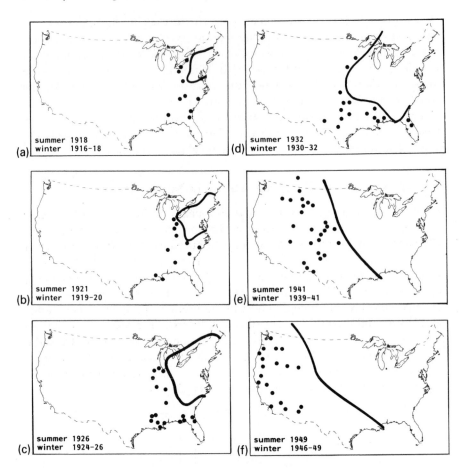

Fig. 4.8 Geographical expansion of the European starling in the United States. The dots are observations in overwintering areas; the drawn line broadly indicates the limit of the breeding range (after Wing, 1943) (see also Robbins, 1973).

Apart from its subsequent range margins, Fig. 4.8 also shows its overwintering locations. These locations, mainly containing young birds that had not previously occupied territories, serve as bridgeheads from where further neighbourhood expansion occurs (Kessel, 1953). Locally, population increase is at first exponential and later assumes a more stable level as shown by Fig. 4.10 for the longest time series, that of New York. Plotted on logistic probability paper, the increase rate is almost linear, suggesting logistic growth. Data from Ohio and Massachusetts show a similar growth type, the fluctuation levels being comparable although their amplitudes are smaller in Ohio than in Massachusetts. The growth rates in these three cases vary; they are lowest in New York and highest in Ohio and Massachusetts,

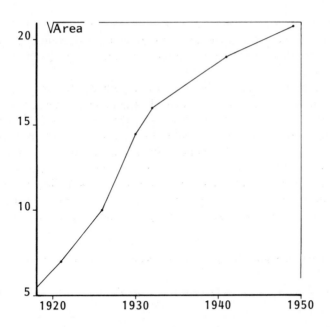

Fig. 4.9 Expansion rate of the European starling in North America (after Kendeigh, 1974).

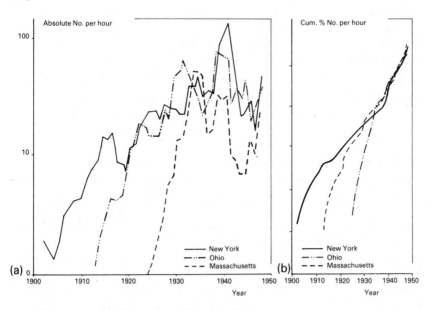

Fig. 4.10 Population growth of the European starling in New York, Ohio, and Massachusetts, expressed in absolute numbers seen per hour and as a cumulative distribution plotted on logistic probability paper (partly after Davis, 1950).

the difference probably being due to absence of immigration in the founder population of New York, whereas immigrants from surrounding regions can add to population growth in the latter two states (Van den Bosch, personal communication).

Dobson and May (1986) discuss why, together with the house sparrow, the starling is the most successful North American invader. One possibility, which may be more general in invaders, is that in their new homeland they suffer from only some of their natural enemies (cf. also Simberloff, 1986), the others having been left behind in their home countries. Of the starling's parasites, roughly one third of the European genera and species also occur in America, whereas half the genera and species the starling has in America also occur in Europe, the other half having been acquired in America (Table 4.1(a)). The same holds for the European house sparrow (Table 4.1(b)),

Table 4.1 (a) Helminth parasites of European starlings in Europe and in North America

| Helminth group | Number of Genera (Species) | | | |
	In Europe	*In North America*	*Total Number*	*Number in common*
Trematoda	17 (26)	4 (4)	18 (28)	3 (2)
Cestoda	9 (12)	4 (5)	10 (14)	3 (3)
Nematoda	14 (26)	6 (10)	17 (30)	3 (6)
Acanthocephala	4 (6)	2 (3)	6 (9)	0 (0)
Total	44 (70)	16 (22)	51 (81)	9 (11)

From data in Hair, J.D. and Forrester, D.J. (1970) *Am. Mdld Natur.*, **83**, 555–64.

(b) Ectoparasites of house sparrows in Europe and in North America

| Ectoparasite group | Number of Genera (Species) | | | |
	In Europe	*In North America*	*Total Number*	*Number in common*
Acarina	8 (35)	5 (24)	10 (43)	3 (16)
Mallophaga	8 (18)	4 (9)	9 (22)	3 (5)
Siphonoptera	1 (7)	1 (4)	1 (9)	1 (2)
Total	17 (60)	10 (37)	20 (74)	7 (23)

From data in Brown, N.S. and Wilson, G.I. (1975) *Am. Mdld Natur.*, **94**, 154–65.

suggesting the reliability of the data. However, exact data on the burden parasites exert on starling numbers either in Eurasia or in North America are not known.

Thus, the starling expanded at a regular rate into North America through nuclei ahead of the wave front. Local populations were growing roughly logistically to mutually comparable fluctuation levels. The burden from parasites was less insofar as this concerns the number of taxa; the ecological burden is unknown.

4.6 Conclusions

Depending on species properties, seasonality and environmental instability, several types of population growth can be conceived. The simplest of these is unlimited exponential growth. Because lack of growth limitation is unrealistic, the logistic equation has been formulated, which solves the problem heuristically, but depends on other unrealistic assumptions. As long as population growth and limitation have not been formulated in mechanistic terms, the logistic equation serves as the simplest description of demographic growth processes. Yet it is unrealistic to take it as a starting point for formulating other concepts, such as the environment's carrying capacity or processes such as prevention of species from trespassing, as this introduces circular reasoning.

Below, I will treat logistic growth heuristically as an empirical description of processes determined either by density-dependent factors or by density-independent ones.

Five

Diffusion

Diffusion essentially results from undirected movements of particles in a homogeneous medium; its direction is brought about by spatial differences in density or concentration of diffusing particles. Thus, steady flows result from locations of high to low density because of the number of particles moving into low density locations is larger than that moving in the opposite direction. This steady flow of particles is a chance of stochastic process described by probability theory. The steadiness of the flow depends on the number of particles; the flow can be described in deterministic terms when chance variation is negligible.

Yet processes that are purely stochastic on one particular scale can be deterministic on others. For example nuclear decay is stochastic at the scale of individual atoms, but deterministic on broader scales where many atoms are involved. Similarly, population genetic change is stochastic in small populations during short time periods and then known as genetic drift. For large populations during similarly short periods, genetic change, though depending on the chance mechanism of recombination, is deterministic. But on yet longer time-scales, stochasticity of climatic change plays its part, causing the process to be stochastic again, even for very large numbers of individuals (Lande, 1976). Whether a process can best be described in deterministic or stochastic terms therefore depends on the number of particles involved and on the choice of spatio-temporal scale of variation.

So far, I have described spatial diffusion by using the general term particles, comprising both inanimate and animate dispersal units or propagules. But diffusion of animate propagules differs in several ways from inanimate particles. Firstly, the dispersal capacity of propagules varies both between and within species, as does their response to local conditions after arrival. Not only do these conditions vary between sites, but also over time within sites. These sites vary in size, quality and mutual distances and are either randomly dispersed or clumped. These two types of site dispersion can be described using probability distributions, such as the Poisson distribution or the negative binomial distribution respectively and the inter-site distances with the according interval distribution. Thus, the characteristics of a species' dispersal capacity should match those of the interval probability distribution of the inter-site distances related to the spatial site distribution. If they do not match diffusion cannot occur; if they do to only a certain extent the diffusion process is slowed down. Similarly, species require a certain minimum living area for their maintenance implying that their requirements should match the frequency distribution of site sizes. Also, their catchment relates to the size of the site implying that the species' dispersal characteristics should also match the latter frequency distribution. Often catchment probability is enhanced by the propagule being able to identify the site and to drop there. Finally, when environmental conditions vary within sites they can only be temporally suitable in which case the species' properties should match the temporal probability distribution of their suitability.

It is clear, therefore, that the dispersal process is complicated, consisting of several individual processes whose nature is often stochastic. This, in fact, may have led Pielou (1969, 1977, 1979) to equate dispersal with diffusion as seems to be the custom in the humanities (e.g. Brown (1981) in economics; Cliff *et al.* (1981) in human epidemiology; Hägerstrand (1969) in sociology; Haggett *et al.* (1977) in human geography and Renfrew (1987) in archaeology and linguistics). As a mathematical term diffusion concerns only one of these processes, being defined as the movement of particles in Brownian fashion. Thus, these particles become normally distributed around their point of origin, any deviation from normality resulting in a process not considered as diffusion. For example, when propagules are caught in the environment, and distance decay due to catchment is exponential, a contagious distribution, the Bessel function, results. This function consists of two superimposed probability distributions, the normal distribution and the exponential distribution. In this case the composite process is not called diffusion but spatial spread. However, as soon as subjects are made explicit by quantification or by mathematical modelling, we have to adjust our terminology accordingly. Therefore, the term diffusion can still be used in its loose sense in other contexts but here I use it in its mathematical, stricter meaning.

Thus, the probability of species invading areas and their subsequent survival can be formulated in terms of probability distributions of ecological conditions occurring. Also the species themselves must possess properties that match the spatio-temporal characteristics of these conditions. In this chapter, I first describe neighbourhood diffusion between adjacent areas, then some probability distributions, after which I discuss stratification of diffusion processes according to dispersal over various distances. This leads to the concept of the dispersion probability field. Finally, I consider models which integrate diffusion processes with growth processes as discussed in the previous chapter.

5.1 Neighbourhood diffusion

Neighbourhood diffusion results from individuals immigrating into adjacent areas. However, whether areas are adjacent or not depends on the spatial scale on which individuals disperse; species with large individuals or those with an efficient dispersal mechanism can bridge unsuitable areas more easily than those with small, or with less efficiently dispersing individuals. The concept of neighbourhood diffusion thus applies to intra-species comparisons only. Whereas neighbourhood diffusion is defined for short-distance migration, long-distance dispersal results from jump dispersal (Pielou, 1979); between these two extremes there is a continuum of interme-diate distances (Van der Plank, 1967).

Neighbourhood diffusion, in general, progresses with a more or less closed front, in contrast to long-distance dispersal, which progresses more patchily and with broken fronts if any (e.g. Mollison, 1977). This is because a minimum number of individuals is needed to maintain a certain minimum density in a closed front which decreases geometrically with distance from

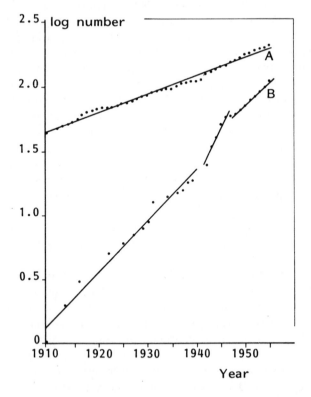

Fig. 5.1 Population growth of (A) the gallant soldier, *Galinsoga parviflora*, and (B) the shaggy soldier, *G. ciliata*, in Britain (after Lacey, 1957).

the species' introduction point. Thus, long-distance dispersal progresses patchily because of shortage of individuals to fill in the gaps, whereas in neighbourhood diffusion there are usually enough individuals to keep up a minimum density throughout the newly available area.

Figure 5.1 shows the spread of two plant species, the gallant soldier *Galinsoga parviflora* and the shaggy soldier *G. ciliata*, in Britain during the first half of this century. First, the rate of spread is geometric in both instances, and despite their introduction at several locations, their progress is roughly continuous. Secondly, *G. ciliata* has spread at a much higher rate than *G. parviflora*. Thirdly, during the war when *G. ciliata* may have taken advantage of the abundance of open, bombed areas, it invaded Britain at an even higher rate (Lacey, 1957; Salisbury, 1943). Thus, despite its apparently

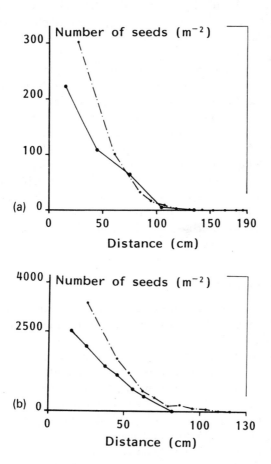

Fig. 5.2 Number of seeds at various distances from the parent plant of (a) the small scabious, *Scabiosa columbosa,* and (b) the Chiltern gentian, *Gentianella germanica* (after Verkaar *et al.*, 1983).

more efficient dispersal than *G. parviflora*, the spread of *G. ciliata* was restricted by habitat shortage at sufficiently short distances. The slower, though steady progression of *G. parviflora*, not being enhanced during the war, suggests that its own dispersal properties, rather than habitat availability, have been limiting. Unfortunately, neither Salisbury (1943) nor Lacey (1957) indicate exactly which properties might explain these differences in spread.

Verkaar et al. (1983) analysed propagule dispersion from parent plants using experimental information on seed dispersal. Using equations for upward acceleration and fall of seeds they simulated the relationship between the parent plant and the distance of seed deposition. Fig. 5.2(a) and (b) show that expected and observed values fit reasonably well, the expected values systematically reaching too low, although following the main observed trends. Moreover, at various heights, taller plants grow in dense tufts and disseminate at greater distances from open vegetation than smaller ones.

Two dispersal morphs occur in many insect and plant taxa, allowing for short-distance and long-distance dispersal respectively (Rose, 1978; Southwood, 1962). In plants, this dimorphism can be expressed in differences in seed size, for example, and in insects in wing length; wing dimorphism can either be genetically determined or be modified by external factors (Harrison, 1980). According to, for example, Southwood (1962; cf. also Duviard, 1977), the long-winged or macropterous form would be functional in colonizing new areas and is found particularly in unstable environments. The short-winged or brachypterous form occurs mainly in more stable environments and allows for population maintenance in wing-dimorphic species. The dispersal process resulting from dispersal by either of these types is then sharply stratified.

5.2 Stratified diffusion

To explain the post-glacial northward range expansion of oak in Britain, Skellam (1951) assumed that every parent oak produced 9 million mature offspring during its life-time. Such huge numbers would be possible because of low competition pressure at the wave periphery. In his model, oaks do not produce acorns until 60 or 70 years after germination, after which they reproduce for several hundred years. Thus, if they would on average, reproduce for 500 years, each would give rise to 18 000 reproductive daughter oaks annually. Okubo (1980) assumes the average number of mature daughter oaks produced by a single tree to be > 9 million for 300 generations! However, such outrageously high numbers, plus those that do not reproduce, would soon compete intra-specifically. Intuitively, neighbourhood diffusion envisaged for the oak's post-glacial invasion into Britain does not suffice.

Table 5.1 Number of settlements (from Cliff *et al.*, 1981)

Time periods	Neighbourhood effect		Hierarchic effect		Combined n/h effect	
	abs.	*%*	*abs.*	*%*	*abs.*	*%*
t_1	1	33	1	33	1	33
t_2	7	41	3	56	9	63
t_3	13	58	9	78	27	100
t_4	19	85	27	100		
t_5	26	98				
t_6	27	100				

Many ways of dispersal often occur side-by-side within a species, thus stratifying the diffusion process. Moreover, combinations of various types of dispersal will lead to the most rapid spread (Van der Plank, 1967) and seems to be the rule rather than the exception in biological invasions. Nuclei of isolated colonizations at large distances from the parent population can enhance spatial spread considerably. Haggett *et al.* (1977) simulated neighbourhood diffusion and compared its rate with that of diffusion along a hierarchy of town sizes by jump dispersal. The diffusion rate of this type of dispersal is higher than that of neighbourhood diffusion (Table 5.1), whereas a combination of them is highest, taking only half the number of steps required by neighbourhood diffusion. In fact neighbourhood diffusion will relate to, for example, the onset of a disease and distance, whereas it would not relate to town size. Conversely, hierarchical diffusion will relate to the arrival of diseases and town size but not to distance, whereas combinations of the two will relate to both distance and town size.

This expectation holds true in Cliff *et al.*'s (1981) reanalysis of Pyle's (1969) data on three epidemic waves of cholera in North America during the last century. In 1832, before the size differentiation of American towns and before the construction of an efficient transport system, cholera dispersed according to neighbourhood diffusion. Later, in 1866, cholera spread according to both types of diffusion, town size possibly having the largest effect. During the intermediate epidemic of 1848, effects of the two types were mixed for the New Orleans route and mixed hierarchical for the New York route (see Chapter 2 and Fig 2.8). Thus, both types of diffusion occur in cholera, either separately or in combination, the resulting hierarchical diffusion process being stratified to both town size and to distance between the towns. Moreover, their relative proportions of variance explained are not stable but vary with time.

Similar processes can occur more generally in biological invasions. For example, Mack (1981) documented the spread of cheat grass, *Bromus*

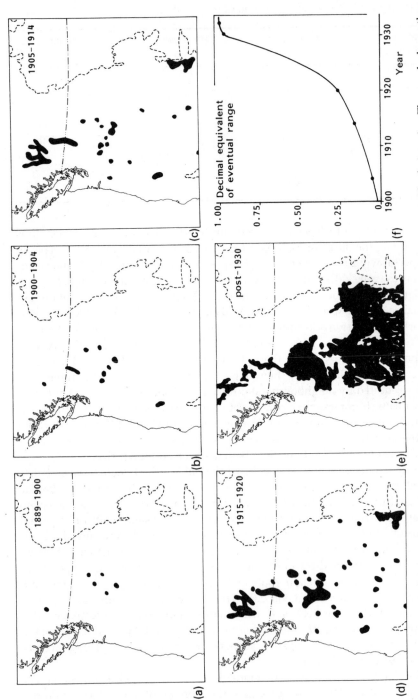

Fig. 5.3 Geographical expansion of cheat grass (drooping brome), *Bromus tectorum*, in western North America. The graph shows the expansion rate in percentages of the eventual range (after Mack, 1981).

tectorum, into western North America (Fig. 5.3). Between 1882 and 1900 it was found at scattered locations only, after which both the number of locations and their sizes increased. At present it is a sole dominant in ranges and fields (Mack, 1986). Apart from its distribution, Fig. 5.3 also shows the increase in rate of cheat grass as a proportion of its present status, suggesting logistic growth (Mack, 1981).

Mack (1985) reasoned that the number of foci is more important than their individual size because the number of propagules arriving in uninvaded areas from smaller foci is greater than that from larger ones. From simulated data, Fig. 5.4 shows this for various focus numbers with equal total surface areas by using the equation

$$A = [r^2 + 2V\sqrt{n}\,(rat) + na^2t^2]$$

where A = area, n = locus number, a = growth rate = 0.05, r = focus radius with a maximum of 1 for the largest, single focus, and t = time.

In wind dispersal yet another type of stratification can be recognized. Johnson (1957) regressed log density of insects against log height in the air,

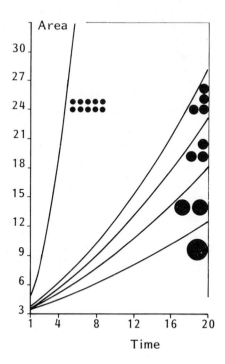

Fig. 5.4 Expansion rate as a function of the number of foci with equal total surface area (after Mack, 1985).

and obtained rather straight lines. Moreover, the regression coefficient varies with weather and day time. Using this formula and observations at various heights, the proportion of the population occurring at various altitudes at various times during the day could be inferred; on a particular day between 11.00 and 12.00 hours, 50% of the population of the fly *Oscinella frit* (L.) would have occurred above 400 metres (Table 5.2). High-speed winds in higher atmospheric strata can then transport large proportions of the populations over considerable distances. Thus, apart from locally short-distance migration just above the vegetation, long-distance migration also occurs in many species, depending on temperature and air turbulence. Where the insects, or propagules in general, drop is uncertain, as this depends on wind direction and velocity as the most important factors.

When various processes operate at different scales of variation the resulting expansion wave will not be regular, rather, its velocity varies over time. As another explanation from those of genetical and physiological or community adjustment after settlement (Chapter 3), at low initial population densities the percentage of long-distance migrants will not add significantly to wave velocity. This velocity then depends entirely on neighbourhood diffusion. But as density in the founder population increases the absolute number of migrants settling at long distances away from it increases as well, eventually dominating wave velocity. This velocity increases gradually, depending on the rate at which the ratio of the two processes, operating on the two spatial scales, shifts. Thus, the possibility that initial population

Table 5.2 Estimated numbers of *Oscinella frit* (L.) above an area (1,000 feet^2) to a height of 3000 feet at different times of the day (from Johnson, Taylor and Southwood, 1962)

Hours G.M.T.	Total frit flies (thousands)	Percentage of population higher than 20 feet	Approximate median height of flight in feet
06–07	4	4	<5
07–08	7	34	<5
08–09	26	96	550
09–10	32	92	800
10–11	45	91	1,150
11–12	92	89	1,300
12–13	47	87	700
13–14	11	78	550
14–15	3	56	25
15–16	6	71	125
16–17	8	92	500
17–18	19	3	<5

growth is slow because of physiological or community adjustment is not the only explanation. The idea that various processes exist that operate at different scales thus simplifies our explanation of deviations from non-linear wave propagation.

If two dispersal types occur together in higher taxa the spreading process is stratified at a higher taxonomic level and leads to disharmonic biotas (MacArthur and Wilson, 1967). One of the long-distance mechanisms in plants, for example, is zoochory by means of bats and birds, whereas dispersal by water and wind are short-distance mechanisms (cf. also Nip-van der Voort *et al.*, 1977). This discrepancy can be understood by assuming that dispersal by birds and bats is directional and water and wind dispersal non-directional. Directional dispersal densities are inversely proportional to the distance from point of release, whereas non-directional densities fall off against the distance squared (Elseth and Baumgardner, 1981). Probability distributions describing other dispersal processes are also exponential with different negative exponents. The smaller density decrease with distance of zoochorous dispersal enhances the probability of arriving in sufficiently large numbers at larger distances. When the exponent is -2, the chance distribution of arriving at various distances from the origin is Gaussian. However, other probability distributions can give the same exponential decrease.

5.3 Some probability distributions

A probability distribution widely used in genetics describes the chance that two alleles of a gene combine in the next and the following generations. Here, the chances of allele A combining with the same allele A or with another one, a, are supposed to be equal. Then, in the next generation we expect one combination AA, one Aa + one aA = two Aa's, and one aa. For N individuals, we will find one quarter N AA, half N Aa, and one quarter N aa, which will be constant over the generations without effects of chance variation (genetic drift), mutation and selection. This probability distribution is known as the binomial distribution because, in this example, two alleles are involved. In the case of a spatial process, we can think of propagules either stopping (p) or continuing ($q=1 - p$), or going right (p) or left (q). The equation can be expanded from two to more than two alleles — or, in general, to more than two alternative events whose probabilities of occurrence are $p, q, r, \ldots$, resulting in a multinomial distribution. Moreover, p and q can form more complex combinations, when, for example, one allele p combines with two alleles q, etc. In the above case, two combinations k are possible, resulting in $(p + q)^2 = 1\ p^2 + 2pq + 1\ q^2$; for a large exponent k, in practice the resulting frequency distribution becomes smooth, although theoretically it is still discrete. If so, it can be approached by the normal or Gaussian distribution, which is less tedious to compute than the binomial with large k. For large N the outcome is deterministic although it results from a chance or stochastic process. For its application we make two assumptions; that the two alternative events p and q are equal, and that they are discrete. This second assumption implies that the binomial distribution belongs to the family of discrete probability distributions. The Gaussian distribution belongs to another family, that of the continuous distributions.

The Poisson distribution is another probability distribution of wide applicability. Although different, one can approach it by making p of the binomial very small, say 0.0001, and its complement q, consequently, large, in this case 0.9999. Apart from this, k should also be large. As an approximation, the Poisson distribution describes the chance of rare events p happening. In spatial processes, it describes, for example, the number of seeds falling on a certain surface area, and in temporal processes the occurrence of occasional events happening during a certain period. In such instances, one partitions the area or period into a number of equal blocks or time units and counts the rare events. From the number of times the event occurred only once, twice, three times, etc. in the various units, a frequency distribution is constructed. For, on average, many rare events happening, this distribution is skewed to the left, or even L-shaped, and it is more symmetric for larger averages (Fig. 5.5).

Apart from assuming that p is small, one also assumes that the chance of falling somewhere in the square, or of happening in a given time period, is the same for all blocks or time units. If not, the spatial or temporal dispersion either becomes clumped when too many events occur together, thus

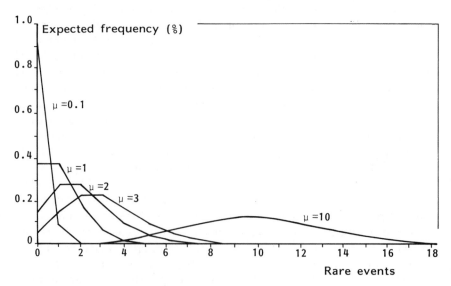

Fig. 5.5 Frequency polygons of the Poisson distribution for various mean values, μ.

leaving too much open space in between, or uniform or regular when too few events occur together, the events then being evenly spaced. Since in the Poisson distribution the mean λ equals the variance σ^2, the ratio λ / σ^2 is an index of dispersion, giving values >1 for clumped dispersions and <1 for uniform ones. However, this does not imply that for $\lambda = \sigma^2$ the distribution is necessarily Poisson; other distributions resulting from different processes can give the same equality.

Other biologically interpretable models are the log-normal distribution and the log-Poisson distribution (cf. Cassie, 1962). These pertain to those distributions that are normally- or Poisson-distributed on semi-logarithmic paper respectively, implying that the exponent of some biological variable is normally- or Poisson-distributed.

In invasions, or in temporal dynamics and gap dynamics, the interval between adjacent or successive events is of interest as both have to be bridged. In time, such intervals can be bridged by a seed bank, for example, and in space by dispersal. For Poisson-distributed spatial or temporal events, the mean interval to be bridged together with the standard deviation is $1/\lambda$. For many cases taken together, the Central Limit Theorem of probability theory states that their means λ_i are normally- or Gaussian-distributed. Thus, favourable sites may be Poisson-distributed with local means λ_i, whereas as seen on another broader, spatial scale, the various local values λ_i are normally-distributed. The process as a whole is then stratified at two levels of variation.

The offspring of Poisson-distributed individuals can themselves be Poisson-distributed around their parents resulting in a clumped pattern. This distribution, the Neyman-type A distribution, is also skewed to the left, though different from the original Poisson distribution. It is called a contagious distribution as it is composed of two superimposed distributions (e.g. Pielou, 1977). Another contagious probability distribution is the negative binomial, so called because the formula of the binomial distribution $(p + q)^k$ is altered into $(p - q)^k$ is related to population structure. Although this distribution has often been applied to describe clumped distributions, it has no biological interpretation in terms of a generative process. One of its applications is the logarithmic-series distribution, which describes the abundance distribution of the species in a sample (Fisher *et al.*, 1943; Pielou, 1975) and is an approximation of an interpretable model based on a sampling process (Hengeveld and Stam, 1978). Other, biologically interpretable contagious distributions used in the context of spatial analysis are the Bessel function and the double exponential distribution, to be discussed in Chapter 7 (see Hastings and Peacock, 1975 for a practical overview of the commonest distributions).

An empirical example of probability distributions characterizing a dynamic field situation concerns the spread of ground beetles from Sweden across the Aland archipelago in the Baltic. Ås (1984) compared the relative

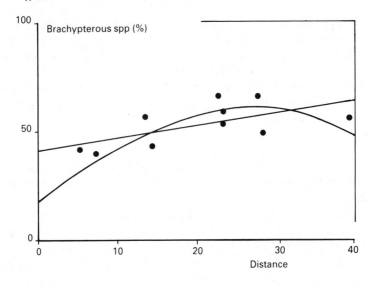

Fig. 5.6 Relationship between the percentage of brachypterous ground-beetle species in ten Baltic islands and the distance of the island from the Swedish mainland. Dots, observations; drawn line expected percentage (after Ås, 1984).

efficiency of non-directional dispersal by wind and water, assuming exponential decay and normal diffusion for these mechanisms, respectively. Among the species sampled, two types can be distinguished, winged or macropterous species and unwinged or brachypterous species, which in that area disperse by wind and by water, respectively. Assuming that mean dispersal distances differ among these two types, being 5km for the unwinged water dispersers and 10km for the winged air dispersers, the proportion of unwinged species can be calculated as a function of distance from the Swedish mainland. Figure 5.6(a) shows that the expected proportion of unwinged species first increases with distance and then drops. This expectation accords with observed values on the islands (Fig. 5.6(b)). It is reasonable to assume that the same stratification also occurs between the genetic morphs within wing dimorphic insect species.

5.4 The dispersion probability field

The probability of moving a particular distance during a certain time, say one generation, is central to considerations of diffusion rates. This probability is estimated from the frequency at which various distances from the source area are covered (Fig. 5.7). In historical data one counts the number of marriages at various distances from the couple's birth places. However, this procedure is time-consuming if it can be applied at all for slow processes and for those covering large distances. Yet Wolfenbarger (1975) collected

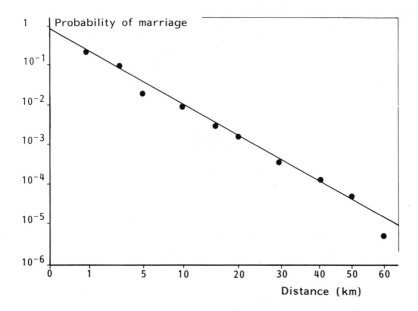

Fig. 5.7 Probability of marriage as a function of distance between village of origin in northern Italy (after Cavalli-Sforza, 1958).

many instances from the literature that all indicate exponential decrease with increasing distance. The area encompassed is called the neighbourhood area in genetics, the contact area in epidemiology, or the information field in human sociology. Hägerstrand (1969) called the probability distribution derived from such frequency distributions the information probability field (cf. Brown, 1981, using 'mean information field'). Similarly, for biological processes, the general term dispersion probability field can be used encompassing that area where a defined proportion of invaders — be they real invaders, genes, diseases, etc. — will be found. It describes the decay of this proportion with distance. This field is rotationally symmetric when the propagules disperse, on average, the same distance and at the same rate in all directions, and asymmetric when they do not (Fig. 5.8). When dispersal direction is discarded one implicitly assumes rotational symmetry.

The decrease of the proportion of propagules dropping with increasing distance from the source area always follows an exponential distribution, although the exponent is particular for the case at hand. The form this exponential decrease takes is

$$\log(G(x,t)) = -[(x-v)/a]^b$$

where $G(x,t)$ is the proportion of propagules having travelled distance x at time t, v expresses the degree to which the propagules moved in a certain

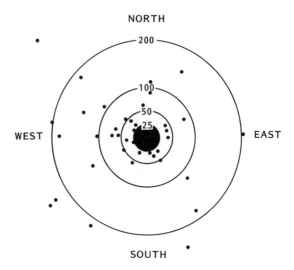

Fig. 5.8 Relationship between birth place and breeding locality of American robins, *Turdus migratorius* (after Kendeigh, 1974).

direction, and *a* is the mean distance covered. The parameters *a* and *v* are time-dependent; without drift occurring, *v* equals zero. In this equation, *b* is the exponent of interest, determining the rate of decay of the proportion of propagules with distance. This parameter can adopt various values depending on the chance process happening. For exponential decay, for example, *b* = 1, whereas *b* = 2 for normally-distributed patterns. Of course other patterns are also possible resulting from other processes and giving different values of *b*. However, exactly because several processes can result in the same value of *b*, and because of the possibility of errors in estimation, backward reasoning is hazardous: it cannot be concluded from, for example, *b* = 2, that the chance process was Gaussian (cf. also Ito and Miyasita, 1965).

For example, Cavalli-Sforza (1958) counted the number of marriages at various distances from the couple's birth places and obtained a highly skewed, L-shaped frequency distribution (Fig. 5.9(a)). To compare the shape of this distribution with the theoretical one, he plotted the probability distribution for normal diffusion in the same figure. As these shapes are dissimilar (Fig. 5.9(a)), he concluded that the mobility is not the same for all individuals but that it varies. Assuming varying mobility gives more similar distributions (Fig. 5.9(a)), although the theoretical one is still humped rather than L-shaped. As he could not construct an analytical model giving the same shape, Cavalli-Sforza fitted an empirical function $F = c.e^{-kVx}$, where $c = k^2/2$. This equation, therefore, contains the distance parameter x only, k being the scaling factor. Thus, a linear frequency decrease is expected on semi-logarithmic paper which even improves when, instead of frequencies,

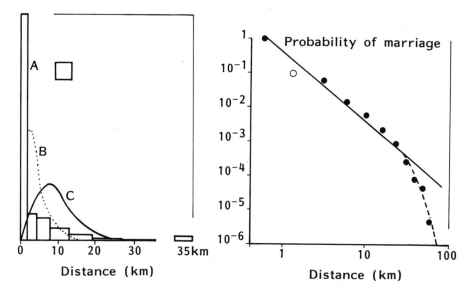

Fig. 5.9 Observed frequency distribution of distances between village of origin of mates and the expected probability according to various models (after Cavalli-Sforza, 1958).

the y-axis represents probability of marriage (Fig. 5.9(b)). These probabilities, in turn, can be calculated in two related ways of which the one expressing them as percentages of the number of people at the distance concerned seems the best.

Thus, Cavalli-Sforza's (1958) empirical equation allows for distance only as a possible explanation of the various probabilities and is specific for his data. As a measure of variation of the individual parish values, the range of variation has been added to the mean values in Fig. 5.9(b), larger numbers of people attracting newly married couples. Then, the probability of marriage would be directly proportional to the number of people in the respective villages, as well as inversely to the distance squared. Fig. 5.9(b) shows a good fit for the shorter distances, but it breaks down for the larger ones, although these together contribute less than 5% of the total number of marriages.

Dispersion probability fields can, therefore, be constructed in various ways, each of them making different assumptions as to the underlying processes. Sometimes only a single process is involved as in normal diffusion, whereas in other cases two or more processes account for the variation found, as in the effect of distance and village size on the probability of marriage. When there are two or more processes involved, the diffusion process can be stratified accordingly; when the same process repeats itself at various scales of variation and runs from high to low densities, a particular type of stratified diffusion is recognized, called hierarchical diffusion (Cliff *et al.*, 1981). They are all types of the same spatial phenomenon of biological invasions.

5.5 Diffusion and population growth: the advancing-wave model

When a moving particle either stops (p) or continues moving (q) and when $p = q = 1/2$, the probability that it continues in n movements is binomially distributed, this may be approximated by the normal distribution for large n. The dispersion around the liberation point or the mean of the normal distribution can be characterized by the variance (σ^2), or the standard deviation (σ), and the dispersal rate in terms of changes in the variance or the

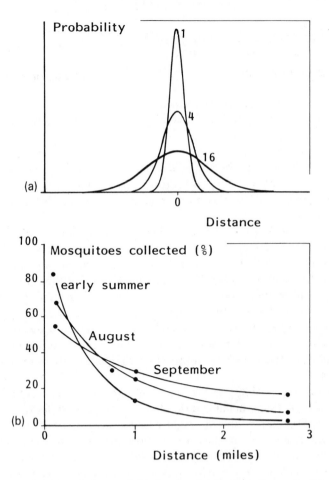

Fig. 5.10 Expected (a) and observed (b) density distribution after spreading from an invasion centre (Figure 5.10(b) after Wolfenbarger, 1975).

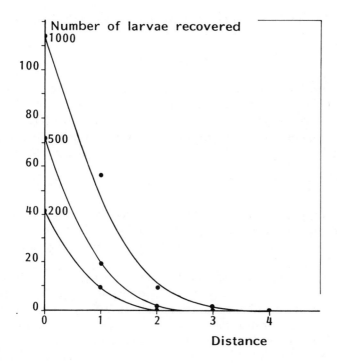

Fig. 5.11 Number of larvae of the corn borer recovered at various distances from source populations of three different sizes. The distance units are rows of corn (after Kendeigh, 1974).

standard deviation over time. After n steps, the standard deviation equals $V \sqrt{n}\, \sigma$. Thus, Von Haartman (1949) calculated for the American robin, *Turdus migratorius*, that after 50 generations with a mean dispersal rate of 16 km, $1 \times \sigma$ or 68% of the offspring occur within a radius of 112 km from the point of origin. Since the number of steps is proportional to time, dispersal rate also increases with the square root of time. However, without reproduction taking place, the mode of the spatial distribution decreases with the square root of time as well (Figs. 5.10(a) and (b) for expected and observed instances). Thus, we can define the diffusion rate as $D = \sigma^2/2\tau$ for distance units σ and time units τ. The mean-square displacement along a transect is then $\langle x^2 \rangle = 2Dt$ and that in a two-dimensional plane $4Dt$. The fraction of the population $1/N$ lying outside a circle with circumference $2\pi r$ is e exp $-r^2/4Dt$, this fraction thus dropping with the distance-squared from the source location or area. In exponential decay, the exponent of the decrease of the fraction of the diffusing population equals one instead of two.

For normal dispersion, the area $A = \pi r^2$ containing a certain fraction of the population is

$$A = 4\pi Dt(\ln N),$$

implying that the area of the boundary circle in this fraction is proportional to the size of the spreading population. This, in turn, means that larger populations tend to expand more rapidly than small ones (Fig. 5.11), and also that dispersal relates to growth rate. When growth rate follows the equation $N_t = N_o e^r t$, or $\ln N_t - \ln N_o = rt$, the expansion rate of the area becomes

$$A = 4\pi Drt^2,$$

if $\ln N_o$ is negligibly small relative to $\ln N_t$. Then the square root of the area occupied — or the radius of the area times the constant $\sqrt{\pi}$ — increases linearly with time, and changes in expansion rate reflect changes in growth rate (Fig. 5.12).

This model, combining spatial diffusion with logistic growth of local populations, is known as the advancing-wave model. So far, it has not been extensively applied to biological data. Outside biology *sensu stricto* applications concern reconstructions of historical processes, such as the spatial spread of genetic traits underlying various human blood groups (e.g. Ammerman and Cavalli-Sforza, 1984), present-day epidemic waves (Van den Bosch *et al.*, 1988 *a, b, and in press*), or epidemic waves in the past (e.g. Cliff *et al.*, 1981; Pyle, 1969).

5.6 General-transport models

Diffusion is just one component of a more complex process of a species' spatial dynamics; it describes processes leading to shifts in range delimitation without considering its internal structure and dynamics. But ranges are internally far from uniform; they are patchy, numerically structured and dynamic, locally reflecting overall ratios of birth and death rates. Statistically, numbers are highest in the range centre and taper towards the margin (Hengeveld and Haeck, 1981, 1982). Thus, in particular years the mode can be acentric, and polymodal ranges may occur containing several density peaks. The location of each mode can shift over time depending on environmental conditions, reproduction capacity, mortality and the species' dispersal capacity. The abundance level at the mode relates positively to range size (Hengeveld and Hogeweg, 1979) and depends on the same parameters. Thus, ranges as uni- or multimodal two-dimensional surfaces can be conceived of as the species' response surface relative to changeable environmental conditions cf. Hengeveld, 1989.

Depending on these conditions, the overall species abundance level can increase, the range can expand or contract in all or in certain directions, or it

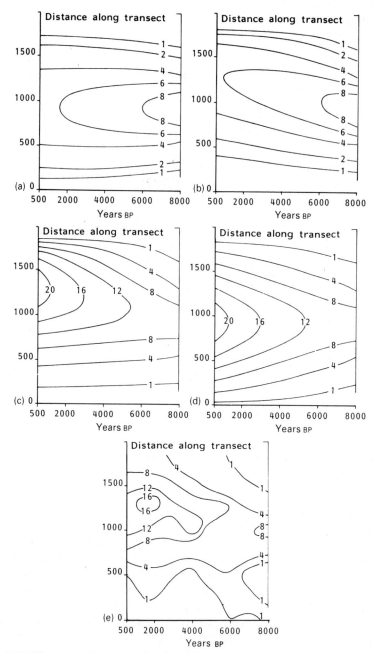

Fig. 5.12 Observed (a) and expected (b) percentages of American beech (*Fagus grandifolia*) pollen along a transect from Tennessee (0) to Quebec (1850km) at various instances in time. The expected values were obtained when only the diffusion term is non-zero (b), when both the diffusion and the advection terms are non-zero (c), when both the diffusion term and the source/sink term are non-zero (d), and when all three terms are non-zero (e) (after Dexter *et al.*, 1987).

Fig. 5.13 Spread of the muskrat from individuals released near Prague in 1905 (after Elton, 1958).

can shift. In general-transport equations outside biology, these three processes are known as the source/sink, the diffusion and the advection terms. For biological data, Dexter *et al.* (1987) applied these equations to changes in the abundance distribution of North American beech during the Holocene. First they selected a transect of 1850 km in eastern North America, known from 300 samples for the last 14000 years. Then they selected a 7500-year period running from 8000-500 BP from this total time span. The resulting chart with percentages of beech pollen present in the samples along the transect and through time was subdivided into an 8 × 8 regular grid (Fig. 5.12(a)). The cells of this grid contain the median percentages as spatio-temporal intensity estimates of the occurrence of beech. From these values the coefficients of the general abundance level, represented by the source/sink term α, the diffusion term D and the advection term V were calculated. Using these coefficients, spatio-temporal charts were simulated for cases where only D, both D and V, both D and α, and

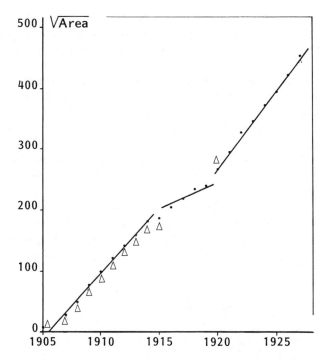

Fig. 5.14 Square root of area occupied by the muskrat in successive years after their liberation near Prague. Triangles: observed area; dots: calculated area, also covering the blank area in Fig. 5.13.

when all three coefficients are kept non-zero, respectively (Fig. 5.12(b-e)). In the first case the range expands, in the second it expands and spreads in a certain direction, in the third it grows, and in the fourth it grows and expands in a certain direction. Only the combination of the three coefficients describes the observed pattern best, accounting for 91% of the total variation. This can be enhanced to 94% by ascribing low values to α between 8000 and 7000 BP and after 500 BP.

Thus, a diffusion parameter only explains part of the total process; ideally it should be supplemented by one of the overall abundance level and by one concerning the direction of spread.

5.7 An example: the muskrat in Europe

The classical example of invasive species behaviour, the European invasion of the muskrat, *Odontra zibethicus*, shows the aspect of diffusion on the broadest spatial scale and that of population growth on the finest.

Apart from many releases all over Finland and the USSR, the muskrat invaded Europe from broadly four liberation points with the oldest one in

1905 near Prague. It is interesting with regard to inbreeding that the many millions of muskrats, today causing great agricultural trouble and ecological cost all over Central Europe, originated from only five individuals released near Prague!

The population took off rapidly expanding its range year after year (Fig. 5.13). Moreover, this range expansion was linear from the beginning (Fig. 5.14); only during and shortly after the first World War did it slacken its rate, coinciding with famines in Eastern Europe at the time (although Nowak (personal communication) considers this coincidence spurious). Fig. 5.13 shows that the range did not expand at the same rate in all directions, but mainly to the southeast and northwest. Also, within the Federal Republic of Germany this rate varied according to topography and to general weather conditions. Schröpfer and Engstfeld (1983) mention that mountain ranges form regional barriers because of the species' preference for marshy and very wet habitats, thus slowing down its progression in Bavaria. Consequently its regional expansion cannot be radial, as it is on the broad Euro-

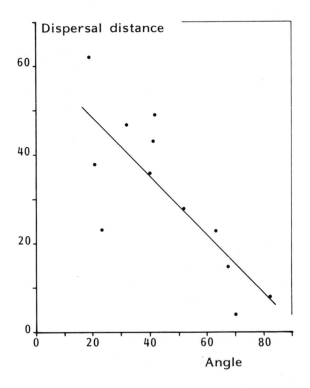

Fig. 5.15 Relationship between dispersal distance of the muskrat and the angle of the direction of its dispersal and the water courses in northern Sweden (after Danell, 1978).

pean scale, but it is star-like, following river valleys. During dry years expansion rate is lower and more linear because of the decrease of wet land. On the other hand, the expansion also slows down in regions with abundant wet habitats, in the extreme north of Western Germany east of the Netherlands and south of Denmark, where the muskrat builds up huge populations at the expense of range expansion. Similarly, the invasion into northern Sweden is hampered by the main direction of local water courses running perpendicular to the direction of the expansion wave (Fig. 5.15).

Range expansion, together with the species' overall numbers, is therefore complicated at this regional scale. Complications are even enhanced because the muskrat entered the country at four locations. First it invaded from the southeast into Bavaria, then from the east along the river Elbe in the north, then in the southwest from liberation points in the Alsace, and finally eastwards from Belgium and the Netherlands in the northwest. Thus, the areal expansion rate is irregular, only becoming logistic in the second half of the expansion period (Fig. 5.16).

Thus, on the broad scale of Fig. 5.13, invasion waves form regularly

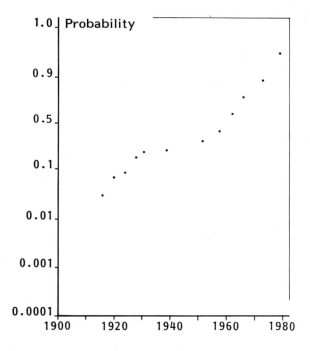

Fig. 5.16 Invasion rate of the muskrat in Germany expressed by the square root of the area covered against time. The y-axis represents the cumulative logistic probability (after data in Schröpfer and Engstfeld, 1983).

progressing, closed, directional fronts, whereas on finer scales they are linear and broken up. Expansion rates and directions are not specific to the species as they depend heavily on regional topographical factors, on local factors and on temporal variation of factors influencing the species directly or indirectly.

At a still finer spatio-temporal scale, Dutch data allow analysis on a yearly basis and at a spatial scale of only tens of kilometers. Initially, after its

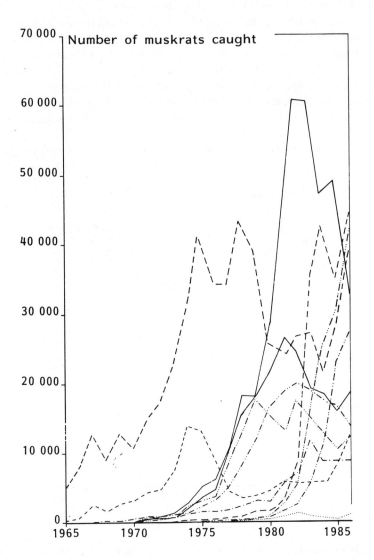

Fig. 5.17 Numbers of muskrats caught in successive years in the Netherlands (after internal reports of the Commission for Muskrat Eradication in the Netherlands).

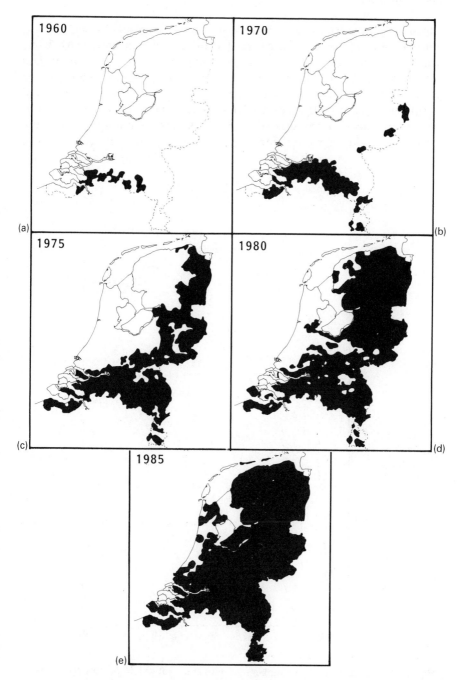

Fig. 5.18 Progression of the muskrat in the Netherlands as shown by its distribution at five-year intervals (after internal reports of the Commission for Muskrat Eradication in the Netherlands).

invasion into the Netherlands in 1941, its numbers remained stable at a low fluctuation level (Fig. 5.17), mainly due to control measures being taken. Due to the shortening of the working week in 1961, however, the muskrat started its rapid increase, thereby extending its range from the south. In the early 1970s it started to invade from the east as well, and then moved northwestwards from the higher, sandy parts of the country into the lower, wetter provinces in the west (Fig. 5.18). Because it damages roads and dykes, an eradication programme was set up. Yet, in spite of catches at

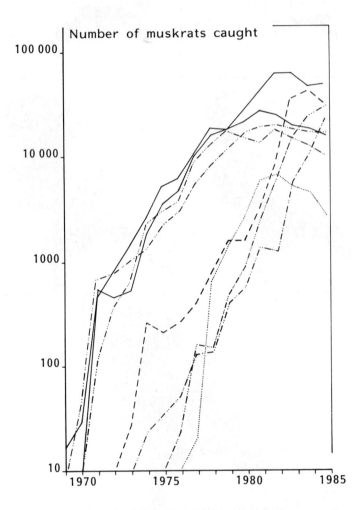

Fig. 5.19 Logarithm of the numbers of muskrats caught in successive years in the Netherlands (after internal reports of the Commission for Muskrat Eradication in the Netherlands). The different curves show data for individual counties in the Netherlands.

present amounting to over 300000 animals annually, its progress seems, though modified, unhampered. The eradication programme keeps track of all the catches in each municipality making provincial counts available.

Figure 5.17 shows the untransformed counts for each of the eleven Dutch provinces. The combined pattern is heterogeneous suggesting that population increase depends on demographic invasion parameters. It also shows that after initial population buildup the provincial populations fluctuated for one or a few years at a high level, after which they dropped to a lower level. These levels vary in height, which will be partly due to differences in total surface area per province and partly to the amount of suitable habitat within each province.

However, reproduction rates cannot be judged easily from this figure, the curves being steeper at higher abundance levels. Figure 5.19 shows the same data, transformed logarithmically for the national scale and the provincial scale. This representation makes the rates of increase comparable both within and between provinces. The slopes of the curves express the

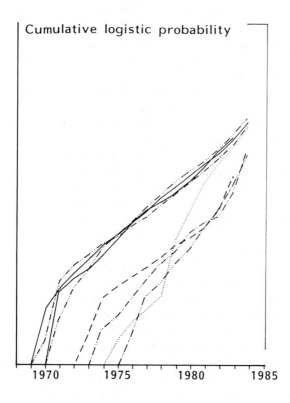

Fig. 5.20 The same as Fig. 5.19; the y-axis represents cumulative logistic probabilities. The different curves show data for individual counties in the Netherlands.

rate of population increase, which, for the provincial data is usually constant over a certain range, after which it levels off or even decreases. The curves run more or less parallel to each other although they reach different levels: they almost all level off at some point and decrease afterwards. Up to the present, the national numbers are still increasing steadily (Fig. 5.19), implying that population decline in certain provinces is still compensated for by increases in others.

Thus, taken as a whole, each provincial curve is more or less S-shaped. The logistic transformation of the original data seems to be justified, making these kind of curves straight, their slope expressing the rate of logistic population increase. The curves of the provincial numbers are straight over more or less their whole range and have similar slopes (Fig. 5.20). Moreover, three bundles of curves can be recognized, those representing the early southern invasion, intermediate ones representing the eastern invasion, and those representing the subsequent invasion of the western part of the Netherlands. One curve, that of North Holland, deviates from the rest, running across the third bundle to the intermediate one. Apparently, it has the greatest logistic growth rate, despite its low absolute numbers and despite its exponential increase rate fall off (Fig. 5.19). The curve for the national numbers is almost straight, showing some slight deviations in the early 1960s. The national population has continued, apparently, to grow logistically during the last couple of years.

Thus, the three representations of the Dutch data each show a different aspect of population growth, particularly at the fine-scale provincial level. The untransformed data emphasize the annual population levels and the independence of fluctuation at the provincial level. These data are interesting both for knowing this species' biotic burden and, hence, for the effects of control measures being taken. The logarithmically transformed data show the exponential increase rates which appear similar at first sight. The various levels reached in each province show that variation is still scale-dependent, the larger provinces or those having most suitable habitats reaching the highest levels. The logistically transformed data show no variation of the provincial growth curves. They express the extent to which the species fits into the provincial or the national habitat space. For different parts of the country, the logistic transformation in particular shows great homogeneity in growth rate of populations started in different years and occurring under different ecological conditions.

The national maps (Fig. 5.18) show a more or less closed front of short-distance spatial progression, though less so than on the still broader scale of Fig. 5.13. Moreover, the national scale of variation showed more or less the same pattern of fluctuation as the provincial level, despite some compensation effects of the various, locally-lagging invasion phases. The various representations thus express different aspects of one single phenomenon and should therefore be used either separately to understand single aspects, or jointly, for understanding the whole process.

Apart from these different growth aspects, the process shows diffusion at a broad geographical scale and population growth at a finer spatial scale. The spatial diffusion process and the temporal growth process are two components of the spatio-temporal process of invasion. Fisher (1937) was the first to develop the advancing-wave model, successfully combining both components.

5.8 Conclusions

Spatial spread can be formulated in terms of the chance process of diffusion of biological propagules comparable to that of inanimate particles. However, the diffusion of propagules is more complex than that of inanimate particles because of effects various biological properties of the species have on the process. Of these, reproduction is the most significant process, preventing the species from reaching too low local densities. Moreover, apart from other factors and processes, it determines the rate of diffusionary spread.

Another, though related, feature of diffusion of biological propagules is the possibility of building bridgeheads far ahead of the wave front, serving as nuclei from where new short-distance diffusion takes place. Thus, the diffusionary process becomes spatially stratified due to dispersal over various distances according to several dispersal means or mechanisms. The probabilities of covering these various distances follow different stochastic processes, the results of which can be described by probability distributions. The total area that potentially can be bridged according to such processes has been termed the species' dispersion probability field. The shape and rate of increase of this field depends partly on the properties of the individuals of the species concerned and partly on ecological factors. So far, these ecological factors, external to the diffusion process as such, have not been discussed. This will be undertaken in Chapter 6.

PART TWO
Applications and interpretation

Part One dealt with the theory of invasion processes in terms of spatial diffusion and population growth. Two components were discussed in detail, that of local population growth and that of radial dispersal from one or more liberation points or nuclei, determined by the spatial transition probabilities between locations. These probabilities vary both among and within species according to the dispersal mechanism. Apart from these intrinsic factors invasions also depend on extrinsic features such as climatic and topographic factors.

Part Two puts the theory into practice. First, Chapter 6 discusses the estimation of diffusion and growth parameters, and the limits environmental factors put to invasions. This is followed by a discussion of simulation of invasions in Chapter 7. Although I have tried to build up the information and concepts gradually, thus making them intelligible, Chapter 7 is an exception. Because of its technical nature, much of it may not be understood by readers not already familiar with the pertaining literature. As it does not add new concepts, it can be omitted. The next two chapters deal with the invasion of the collared dove into Europe and with the spatial epidemiology of rabies in Central Europe, respectively. Chapter 10 puts the theory of Part One and the applications of Part Two into the perspective of ecology and biogeography, and that of nature management. The final chapter summarizes the conclusions reached and formulates work to be done in the future.

Six
Parameter estimation and ecological boundary conditions

Before analysing some invasions in the next two chapters, I discuss parameter estimation regarding invasions, along with the ecological conditions which allow or limit their occurrence.

6.1 Parameter estimation

One component in the process of biological invasions is local population growth. Apart from immigration and emigration growth depends on natality and mortality rates. Therefore, one can either estimate effects of net population growth or of natality and mortality rates on those of spatial expansion.

Natality itself can be partitioned according to various parameters, but one can use the distribution of litter size in mammals and birds over age groups, for example, or spore production per fungus lesion, etc. Part of this next generation settles closely around the parent organism, whereas another part disperses over longer distances. In either case some of the settlers produce the next generation, whereas others die sooner or later without having reproduced. Mortality probabilities for various age classes result when the raw data are expressed as percentages. The percentage of surviving reproducers left from the initial cohort at various time intervals gives the timing of reproduction relative to the survivorship curve. Survivorship curves can take very different shapes and are partly particular to the species. Concave curves indicate that mortality occurs particularly during the early life stages, whereas convex curves occur when it is concentrated in the later ones.

Population growth can either be expressed as exponential or as logistic growth. In fact, the initial phase of logistic growth is exponential as effects of density-dependent factors are still negligible. For expansions feeding from populations close to the wave front, the use of the exponential growth rate will suffice (Diekmann, 1979; Thieme, 1979b). This rate is obtained by plotting the logarithm of population size against time on semi-logarithmic paper; the regression coefficient of this line expresses the population's net

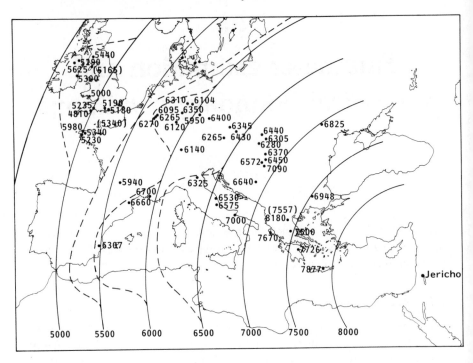

Fig. 6.1 Observed and expected spread of early farming in Europe. The arcs are drawn at 500-year intervals and are centred at Jericho (after Ammerman and Cavalli-Sforza, 1971).

growth rate. But if, as in the simulation of demic expansion of Neolithic man in the next chapter, density-dependent factors cause population growth to be logistic, the regression coefficient of the logistic growth should be used.

The rate of spatial expansion can be expressed by the radial increase of the expanding focus or range (Chapter 4). The square root of the area occupied as a measure of mean radial increase can be taken for irregular shapes of ranges. For discontinuous instead of closed fronts one can draw concentric circles and thus estimate the species' arrival time. Arrival time can either be estimated directly or be derived indirectly from the age of the individuals or populations. Radial distance can then be regressed against arrival time, either for the area between two full concentric circles, or for individual circle segments when expansion rate varies in different directions. Figure 6.1 shows the latter case for the expansion of Neolithic farmers into Europe.

Similarly, the probability of settling at various distances from the source can be estimated by plotting observed numbers of settlers against the logarithm of the distance covered. When the number of settlers at various distances is expressed as the percentage of the total number, the resulting

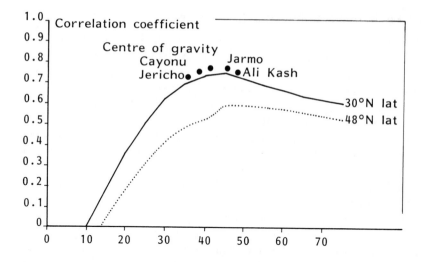

Fig. 6.2 Correlation coefficients between the time of onset of early farming and distance from various assumed diffusion centres. These centres were placed at two latitudes (30° and 48°N) and at several longitudes, differing at 5° intervals. The dots represent correlations with identified ancient towns in the Middle East (after Ammerman and Cavalli-Sforza, 1971).

curve expresses the probability of settling at those distances. The range potentially being covered represents the dispersion probability field (Figs 5.7, 5.8 and 5.9).

So far all measures assumed that the expansion originates from one known source area, focus or origin. But often expansion starts from several regions simultaneously, some of which may even be unknown. For several unknown source areas, one can calculate the spread around a number of regression lines, each assuming a different source area as the expansion centre. For wrongly-chosen areas this spread is large, whereas it is smallest for the proper one. The correlation coefficient between date and distance of several possible centres of spread as a measure of this spread can be plotted against the geographical location of the various potential source areas; the modal value of this curve can be taken to indicate the probable source area (Fig. 6.2). For more source areas, the same, though more complicated, technique can be used.

The following are the principal parameters determining the expansion rates: the mode of introduction of the species (was it introduced or did it escape); the species' demographic and genetic suitability; and whether it is colonizing areas or infecting host populations. As such, they vary as a result of the operation of environmental factors.

6.2 Environmental limitation to invasions

Since the earliest times of applied ecology, attempts have been made to delineate areas where the invasion of an introduced species might stop or where it might progress unhampered. In the main, two approaches have been used which ideally should be combined. The first uses information about the area of origin of the species, supplemented with physiological information. The second uses information on the area invaded.

In one approach, climographs were constructed for various parts in the original species range and in the potentially invasable range. In climographs, a couple of climatic variables, say average temperature and rainfall, are plotted as coordinates for each month separately. This can be done in two ways: all monthly values for several locations in the species' range separately, or per month for several locations together. If the variables chosen are indeed principal ones limiting the species in its region of origin they will also delimit locations in the new, invaded region. Moreover, from additional information on local abundances in the region of origin and from experiments on its physiological performance under combinations of various intensities of these variables, one can predict potential outbreak areas within the new range. This allows us to concentrate control measures in areas of immediate concern.

6.3 Two examples

During the early expansion phase of the alfalfa weevil (*Phytonomus posticus* Gyll.) in North America, Cook (1925) assembled information on the average monthly temperature and rainfall in various parts of the original range. He particularly selected data from areas (1) where it was originally a major pest — around the Mediterranean and in Tashkent, (2) where it was only a minor pest — various countries in Europe surrounding the Mediterranean and the southern USSR, and (3) where it was mentioned only occasionally. From the climographs, plus additional information, optimal conditions are: high temperatures (mean annual temperature $> 10°C$), dry summers (rainfall < 10 cm), relatively high effective winter temperatures ($> -18°C$ without snow, or lower temperatures with thick snow), and a growing season > 150 days. Low winter temperatures and cold, damp spring weather limit adult hibernation, the latter condition affecting the larvae through enhancing fungal growth.

Next, he constructed similar climographs for various parts of the United States allowing him to map potential areas of normal occurrence with severe infestation of alfalfa, of occasional occurrence with periodical infestation parallelling climatic fluctuation, and areas of occasional optimum conditions. Although the species is still expanding, regions with expected and observed severe infestation were the lower valleys of Utah, the Colorado area, western Idaho with the adjacent part of Oregon and western Nevada. In other parts, its abundance varied greatly from year to year.

Thus, according to expectation it was most abundant in the constructed zone of normal occurrence, often injurious in areas of occasional occurrence, and its expansion ceased towards its climatic boundaries. At the time of analysis, it was still spreading into the most favourable areas.

This approach, though powerful, has technical limitations for data processing. Because it is a graphical technique, only two variables can be handled simultaneously for explaining the species' presence or its abundance, three variables resulting in complicated graphs (cf. Hintikka, 1963). However, since the 1950s, computers have been able to reduce vast amounts of data to a few characteristics only. Thus, the number of environmental variables, of observations within or over the years, or the abundance level are no longer restrictive.

Thus, Pimm and Bartell (1980) analysed the invasion of southeastern North America by the fire ant, (*Solenopsus invicta* Buren), using a multivariate statistical technique, Principal Components Analysis (PCA). In PCA the original coordinate system of m axes is replaced by a similar one in which the first axis expresses the main trend or variance in the m-dimensional scatter, the second axis the second largest variance at right angles to the first axis, and so on. Three functions can be distinguished: data reduction, data ordination and ordination interpretation. Data reduction occurs when, from a certain axis onwards, the remaining variance along any further axis is dropped, being considered as noise. Data ordination occurs when one considers the sequence of m points (or m variables) along each axis as an ordered sequence. Finally, ordination interpretation occurs when the ordered sequence is considered to result from a physical factor influencing all variables similarly, though with intensities reflecting the species' responses to it, expressed by their sequences — scores — on each axis. The axis representing a particular trend in the variance is thus interpreted in terms of this measureable, but still hypothetical, physical factor.

Although the fire ant was already introduced in the late 1930s or early 1940s, only the 12 years 1965–1976 were sufficient for statistical treatment. The area concerned was partitioned according to a regular grid with information on both the annual presences, as well as the time series of three weather elements measured monthly for 60 years (occasionally for only 30 years). These elements were mean monthly precipitation, the number of days during which the temperature dropped below 0°C and those where it exceeded 32°C. For 12 monthly characteristics, therefore, the 250 grid squares were characterized by $3 \times 12 = 36$ yearly means. This number was reduced by 3 because during June, July and August the temperature never drops below zero. Thus, instead of two explanatory variables for a limited number of locations, using PCA, 33 variables can be considered simultaneously for 250 locations. Moreover, PCA not only reduces the number of variables; as already mentioned, being an ordination technique, it also allows physical interpretations to be given to directions of variation in the m-dimensional space spanned by the m variables.

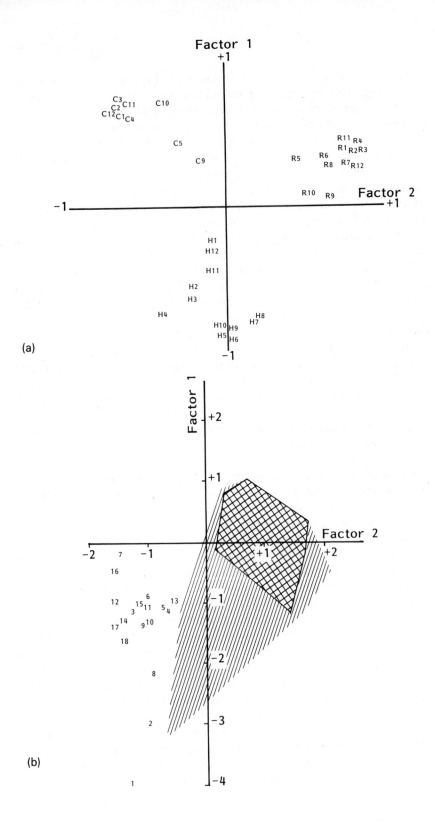

(a)

(b)

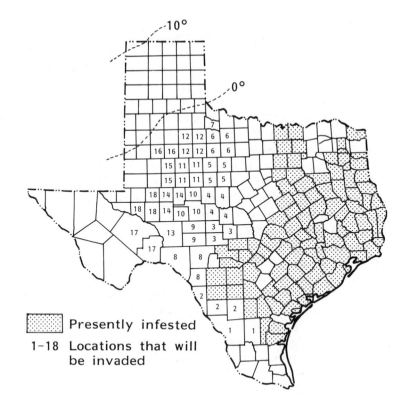

(c)

Fig. 6.3 Ordination of climatic variables (a) and the locations with fire ants (b). (c) gives the observed distribution of the fire ant, together with the locations where it is expected to invade the country in future (after Pimm and Bartell, 1980).

Pimm and Bartell (1980) first ordinated the m (=33) climatic variables measured in each of the 250 locations. Figure 6.3(a) shows the 33 variables from 3 distinct groups of points relative to the first two principal axes. As these axes account for 70% of the total variation in the data, the residual variation on the remaining 31 axes was dropped, thus giving a considerable data reduction without significant information loss. As the mean number of cold days cluster at the positive end of axis 1 and that of the hot days at the negative end, axis 1 was interpreted to represent a temperature gradient. As the mean monthly rainfall clusters at one end of axis 2, this axis is thought to represent the amount of annual precipitation. Next, Pimm and Bartell (1980) plotted in the same coordinate system or factor space, and for each year separately, the locations where the fire ant had been observed (Fig. 6.3(b)). Thus, their first step translates the geographical space in climatic terms, resulting in a climate space. Their second step translates the geographical progression of the ant's invasion into a progression within this climate space. Since it only progressed in one direction within this space, leaving other parts unoccupied, the species gradually exhibits its ecological preference for certain climatic conditions. Finally, Pimm and Bartell (1980) attempted to predict the potentially invadable parts of this climate space, considering the pattern in Fig. 6.3(b). As various points in the unoccupied part of this space represent geographical locations, the potentially invadable parts in southeastern North America can be derived and, hence, the invasion predicted (Fig. 6.3(c)).

Thus, using climatic information and PCA, we can predict the regions the fire ant is likely to invade and where it will stop. Moreover, as the scores on axis 2 plotted against the years do not show a slackening, curvilinear decrease in progression rate, the invasion into the drier parts of the region will progress at the same rate during the next few years. When this rate decreases the species' spatial progression will eventually come to a halt.

This prediction of the progression of the invasion wave and of range delimitation can thus be made in terms of climatic variables. But this does not mean that the variables used are directly causative for this progression and eventual range delimitation; other, correlated variables can be the real, causal factors in this case. So far the relationships are hypothetical. After this phase of hypothesis generation, therefore, one has to add a hypothesis-testing phase in which information on the causal mechanism is assembled. This also applies to other aspects of the analysis of biological invasions, as the next chapter will show.

6.4 Conclusions

For the description and analysis of biological invasions, an array of techniques is available. In this chapter, I discussed several of them, as they appeared in some examples in previous chapters. Examples were given on methods for detecting causative factors that allow species to invade a new region or continent. But all these techniques together constitute the descriptive, hypothesis-generating or inductive phase of analysis, to mention only a few terms in use. This hypothesis-generating phase must be followed by a hypothesis-testing phase. In cases where physical testing is not feasible, for example when large spatial and temporal dimensions or many factors are involved, another technique, simulation, can be used. The next two chapters contain discussions of such simulations.

Seven
Simulating biological invasions

Although Fisher (1937) had already formulated his model of the spatial spread of advantageous genes in the 1930s, it was not until 14 years later that Skellam (1951) elaborated it and applied it to ecological invasions (cf. also Skellam, 1952, 1973). Yet it was another 35 years before this model was applied in realistic constructions and simulation models of demic expansions of human populations into Neolithic Europe. In their long-term studies, Cavalli-Sforza and co-workers maintained a remarkably thorough methodological scheme, starting with parameter estimation and recently completing it with simulation experiments.

Simulation experiments are applied when effects of combinations of parameters cannot be evaluated experimentally or analytically. Moreover, they are the only way of analysis when parameter values are space- or time-dependent. Often such evaluations are necessary when the relative weights of various simultaneously operating parameters are unknown regarding a particular process. In other cases the spatio-temporal scale on which the process operates makes experimentation possible. This makes it unclear whether the set of parameters concerned is sufficient to describe the process, or not. If not, it should be supplemented with additional parameters, whereas in other cases some of these can be superfluous. In still other cases, the result of a process is more sensitive to values one parameter adopts than to those of others. Therefore parameter estimation should, ideally, be followed either by parameter evaluation by means of experimental and analytical techniques or by simulation. However, in most biological invasions and similar processes experimentation is not feasible because of the breadth of the spatio-temporal scale on which they occur. As analytical models are too cumbersome because of the large number of parameters involved, simulation is the only way out.

For the analysis of human demic expansion into Europe, experimentation is not feasible for three reasons; (1) the long time period of roughly 10 000 years involved, (2) the many complicating factors operating during that period, and (3) the subject concerns man. Analytical models cannot be applied because of the great number of parameters. Only afterwards, now we know the process and the parameters possibly involved, might it be possible to formulate analytical models.

7.1 The immigration of Neolithic farmers into Europe

In their simulation model, Rendine *et al.* (1986) assumed that the Neolithic farmers living together with Mesolithic hunter-gatherers in mixed populations reproduced faster than the latter. Migration of part of the populations of farmers and hunters occurred to each of the four neighbouring points in the grid laid over the geographical area. Newborn individuals in the population were allotted genes by binomial sampling from the farmers and the hunter-gatherers; those of the migrants were chosen randomly from the parent population. Finally, for each generation a sample of hunter-gatherers of the size of a defined acculturation rate was transferred to the farmer population. The grid laid over Europe, Africa and the Middle East was a 25 × 35 lattice with seas and mountains left blank. Each grid square of a total of 840 had a surface area of 156 × 156 km²; in each square, population growth over the 400 generations covering 10000 years was logistic. The number of two-allelic genes was twenty with all gene frequencies exactly 50% at the start of the simulations; due to the Wahlund effect, as is expected of deviant genotypic proportions resulting from mixing of several heterogeneous subpopulations, they varied among the hunter-gatherer populations before the arrival of the farmers. Three waves of genetic advance were simulated in each experiment, progressing from three regions in different parts of the lattice and starting at different times. At the end of the simulation experiments the results of these three successive waves were analysed and compared with the present, observed distribution of gene frequencies in the same geographical area.

One measure of a possible gradient generated during the simulations is the similarity of the gene frequency per grid square and that of the surrounding areas at various distances, which is expressed by the correlation coefficient. In the absence of any gradient, such as before the farmers started to invade, the correlation centres around zero, the spread representing the Wahlund variance. In contrast, a gradient exists when neighbouring areas are more similar than those at larger distances, which is expressed by high, positive correlations at short distances and low, negative ones at large ones. Both expectations hold true, indicating the effect of the expansion waves on the eventual genetic constitution. Because the three waves were independent concerning their time and area of origin and because the genes were assumed to vary independently of each other, the resulting cline is complex.

One object of these experiments was to estimate the rate of advance, given certain dispersal and growth rates of local populations. Figure 7.1 shows

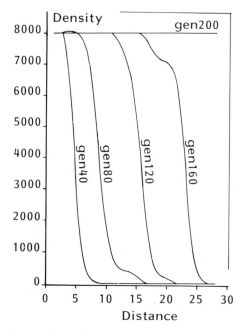

Fig. 7.1 Wave of advance of early farming populations from the onset to the end of the agricultural revolution in Europe, i.e. for the first 200 generations. For every 40 generations, together covering 1000 years, the curves connect densities of local populations at various distances from the centre of origin in the simulation experiment (after Rendine *et al.*, 1986).

the rates of advance for the first wave after various numbers of generations, using parameter values for migration and population growth close to observed ones. For the first and second waves, the theoretically expected rates of advance are in good agreement with those observed. Other experiments by Rendine *et al.* (1986), not discussed here, indicate the possible effects of saturation density and acculturation on the steepness of the spatial gradient.

These analyses all concern aspects of the spatial advance of the wave front of genetic change that could have happened in Europe. Another aspect is that their results cannot be checked using present-day observations. This differs for the resulting geographical pattern of gene frequencies generated, which may not resemble the pattern observed today. Thus, similar to Menozzi *et al.* (1978), who analysed the observed geographical pattern of gene frequencies using PCA, Rendine *et al.* (1986) analysed simulated frequencies with the same technique. Principal axes can distinguish patterns of two-dimensional superimposed clines; the relative values of the scores of grid squares on each of these axes express the steepness of the gradient. Four axes can be extracted from the simulated frequencies. Of these, the first in both the analyses of observed and simulated patterns exhibit the main trend in gene frequencies northwestwards from the Middle East across Europe towards Great Britain and southern Scandinavia. The remaining axes are less easy to compare for the observed and the simulated patterns, the third and the fourth axes of the simulated frequencies possibly representing the expansion waves that originated in two areas. Rendine *et al.* (1986) suggest that the effect of these simulated waves on the pattern of gene frequencies is difficult to distinguish because they travelled in almost the same direction across the lattice.

Thus, by using relatively few parameters, Rendine *et al.* (1986) could simulate the long-term spatial diffusion of cultural change within human populations, partly paralleled and partly caused by biological properties. The properties paralleling cultural change were the frequencies in genes related to human blood groups. Those permitting genetic change involve demographically disadvantageous traits of the early hunter-gatherers relative to the early farmers. Because farming allowed greater families, or even required them (Ammerman and Cavalli-Sforza, 1984), the genes of the Middle Eastern farmers were propagated more easily than those of the hunter-gatherers. But as soon as northwestern European populations had reached the higher density level of the new farming society, the spatial pattern of gene frequencies became fixed.

Thus, the results from these simulations resemble those from observations, suggesting that the parameters chosen explain a significant part of the process that may once have occurred.

7.2 The spread of stripe rust in wheat

Skellam's (1951) model assumes that individuals disperse in Brownian fashion throughout their life time. But this assumption does not always apply; animals and plants often disperse during a certain time, mostly an early phase in their life cycle, after which they settle and start reproducing. Also, Gaussian movement assumes that the individuals disperse in a Brownian way, which does not always apply either. This model thus applies only to those processes occurring at spatio-temporal scales where details of the species' life history or dispersal are negligible. In other cases other models should be chosen.

Van den Bosch *et al.* (1988 a, b, and in press) applied the model developed by Diekmann (1979) and Thieme (1977, 1979 a, b) to experimental observations on the stripe rust (*Puccinia striiformis* Westend) in wheat fields. In their study they aimed at estimating the rate disease fronts propagate after artificial innoculation. Thus, they estimated the rust's gross reproduction (or net reproduction), the time kernel (or reproductive probability density), and the contact distribution (or dispersal density) as the distance decay of the proportion of infectives around the inoculum. As usual, part of the results depend on the properties of the model itself; these should therefore be studied first.

Van den Bosch *et al.* found that wave velocity depends almost logarithmically on net reproduction since the effective distance, as a measure of the distance covered by one generation, increases logarithmically with net reproduction. The steepness of wave fronts also increases with increasing net reproduction (cf. Chapter 9). Its shape is exponential and depends inversely on the variance of the contact distribution. Thus, greater variances characterize flat distributions and, hence, shallow wave slopes. No wave front can be steeper than the slope of the contact distribution as long as this distribution is not Gaussian. For Gaussian distributions there is no such restriction.

Often, the contact distribution results from superimposing effects of two processes, one of which results in the Gaussian distribution and the other in an exponential one. Thus, the distances and directions of movement are normally- or Gaussian-distributed. When fixed proportions of the dispersing propagules are intercepted, the interception rate is exponential. When the effects of both processes are integrated over time, a compound or contagious distribution results, known as a Bessel function, which is more peaked than a Gaussian distribution (cf. Broadbent and Kendall, 1953 and Williams, 1961 for applications to insect distributions). For line sources instead of the point ones discussed here, the double exponential distribution as a marginal one of the Bessel function applies, which was not further investigated. The variance of the Gaussian component, in this context called the diffusion constant, D, increases linearly with wave velocity. This means that shallower contact distributions result in diseases spreading over larger distances. This effect also results from low interception rates.

The reproductive probability density or time kernel consists of three components: latency period, infective period and inoculum production. For this probability density a shifted-Gamma distribution was used as a non-analytical, descriptive model. For long latency periods, low wave velocities are expected, whereas high velocities occur in small mean infective periods and those with large variance. The values any of these parameters adopt depend on environmental factors, implying that effects of these factors should be incorporated in the simulation model of the epidemic or invasion concerned.

This analytical model was applied to two artificial infections, one of which is stripe rust on wheat. Both net reproduction and the contact distribution of this rust were estimated in the field, whereas estimations of the reproductive possibility density were made in the laboratory. For the observed wave velocity (c_{obs}) the disease severity as the fraction of the foliage infected was estimated at increasing distances from several infection centres, summed over eight compass directions. This velocity, being the regression of the size of the patch of infected plants against time, can be compared with the expected wave velocity c_{exp} calculated using the estimated parameter values in the model. This results in comparable values:

$$c_{exp} = 8.0 + 1.5\,\text{cm day}^{-1},$$
$$c_{obs} = 9.4 + 0.8\,\text{cm day}^{-1}.$$

Next, a sensitivity analysis was done to investigate which parameters add most to the variance in the expected wave velocity. Variation in the contact distribution appears to be the most sensitive, that of net-reproduction and the probability density of reproduction being negligible relative to it. This relatively great sensitivity is probably due to the limited sensitivity of the Bessel function.

Thus, the observed wave velocity of the stripe rust can be fairly well simulated by a mathematically analytical model using only a few biologically meaningful parameters.

7.3 Conclusions

With regard to the analysis of invasions, simulation is a powerful technique. For this technique one combines several parameters such that, ideally, they represent the process concerned. Filling in estimates for each parameter gives the expectation about one or more aspects of invasions, such as the velocity rate of wave progression. If the set of parameters in the model is sufficient the calculated rate resembles the observed one; if not, other parameters should be added or should replace those used. However, much depends also on the reliability of parameter estimates, low reliabilities giving poor results.

Usually, models make different assumptions as to the nature of the process as a whole or the parameters to be included. Rendine *et al.* (1986), for example, assumed that population growth is logistic, whereas Van den Bosch *et al.* (1988a, b and in press) assumed that it is exponential. Also, the first authors, using Skellam's (1951) model, implicitly assumed that humans disperse throughout their life time, whereas the model of the second group assumes that dispersal is confined to a restricted phase in the life cycle only. Other model assumptions concern other aspects of the biology of the species analysed. Part of this biological information is inherent to the species, whereas other parts depend on external factors or, technically, on the choice of spatio-temporal scale of analysis. Which model is most appropriate to use in a particular case depends on arguments regarding all of these aspects.

Apart from these arguments for choosing the most suitable model, one can also use a methodological criterion, i.e. whether to use an empirical or an analytical model. This choice often depends on the amount of information available, on the availability of analytical models or on the wish to make explicit assumptions right at the beginning of the analysis or to post-pone their adoption to later phases of research.

Finally, we can distinguish between descriptive and mechanistic models, the second of which give testable predictions. In contrast to descriptive models, mechanistic ones contain parameters on another, independent and usually lower level of integration; the observed value of the phenomenon concerned can then be compared with the expected one obtained from the model. Judging the discrepancy between expected and observed values is essentially testing the applicability of the model. This procedure will be discussed in more detail in Chapter 8 for the European invasion of the collared dove.

Generally it is difficult, if at all possible, to decide once and for all what is a good model, and what is not. Too many arguments are involved in any choice, thereby preventing any such general decision being made.

Eight
Birds invading Europe and America

Our knowledge about biological invasions varies considerably among taxa; the best-known examples usually concern birds. Among these, the invasion of the collared dove is well-known, although other species are often no less spectacular. When taken together common features can be derived that shed some light on the nature of invasions.

This chapter first analyses the invasion of the collared dove in detail; examples are then given of four European and two American invaders.

8.1 The European invasion of the collared dove

At present, so much is known about one classic example of a biological invasion, the collared dove's colonization of Europe, that I will first describe some of the main features of this invasion, then give results of the simulation of its velocity rate, and finally discuss explanations given in the literature.

8.1.1 Description of the expansion

The collared dove, *Streptopelia decaocto,* is one of the best-known examples of a recent natural invader in Europe. Yet surprisingly little is known about its invasion rate and local population buildup; those data that are available had not been analysed fully until recently (Hengeveld, 1988a; Hengeveld and Van den Bosch, in prep.; Van den Bosch *et al.,* in prep.).

During the last century, the collared dove was mainly restricted to the subtropics (Fig. 8.1); in 1900 it was still confined to Turkey, after which it established in the Balkans (Fig. 8.2). There it remained until 1928, after which is started – and still continues – to invade the rest of Europe. Figure 8.3 shows the dove's expansion rate during this century; after 1928 this rate is linear.

Initially, the invasion progressed along three routes (Stresemann and Nowak, 1958); (1) along the Adriatic into northern Italy, (2) north-

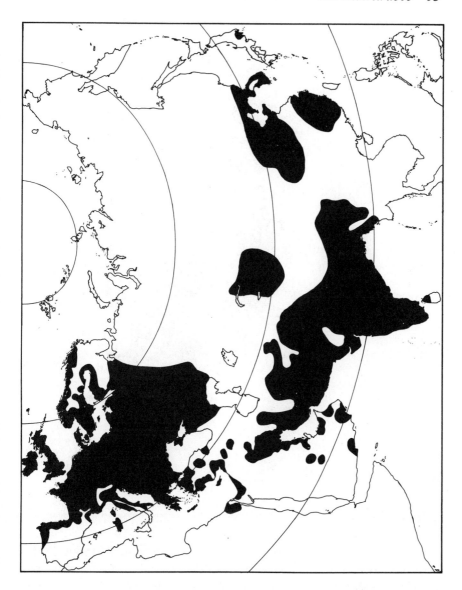

Fig. 8.1 Geographical distribution of the collared dove up to 1985 (after Hengeveld, 1988a).

westwards through Central Europe, and (3) northeastwards into Hungary (Fig. 8.4). Of these, the first and the last routes aborted. At present, after the species reached the North Sea, it expanded sideways, although reaching the sea border need not necessarily be a cause of this lateral expansion. At the same time, since 1939 it has also expanded from other parts of its original

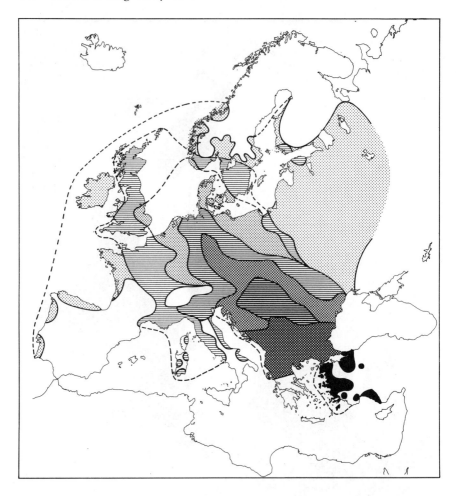

Fig. 8.2 Spread of the collared dove into Europe at 10-year intervals (after Hengeveld, 1988a).

range into Kazakhstan, Turkmenistan and possibly Iran. About 1955 it expanded west of the Jordan Valley into Palestine and since 1979 into Egypt (cf. Hengeveld, 1988a). Thus, its European expansion represents only part of its total range expansion. Within Europe, it mainly followed the valleys between and within the massifs of the Alps and the Carpathians (Fisher, 1953). Towards the north, it is confined to lower altitudinal levels, occurring in India up to 2400 m or occasionally 3000 m, in Switzerland below 650 m before the expansion phase of 1971–1978 and up to 1000 m after it, and in Britain below 300 m. In India it moves altitudinally with the seasons. In Britain it was at first confined to coastal areas before it expanded inland. In the Netherlands, highest densities occur on dry, sandy soil, lowest ones on

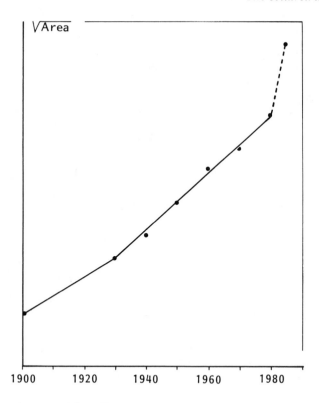

Fig. 8.3 Invasion rate of the collared dove into Europe (after Hengeveld, 1988a).

chalk and loamy soil, and intermediate densities on marine soil (Leys, 1964).

The distribution of dispersal distances can be constructed from a map prepared by Fisher (1953), giving the locations of first observation connected by locations from which they probably derived. As expected, the probability of establishment decreases exponentially with the distance from the parent's nest. Longer distances occur only occasionally, but also follow an exponential distribution, though at a lower probability level (Fig. 8.5). (NB It is, however, possible that this frequency distribution fits a Bessel function; this could not be checked at the time of closing the manuscript.) Therefore, the dispersion probability field does not simply result from a single rate of exponential decrease but it can be stratified. The two strata are that of short-distance, neighbourhood diffusion and that resulting from long-distance dispersal. The establishments at large distances from the source location are often bridgeheads for further expansion far ahead of the wave front. Occasionally they abort but only a few invaders are sufficient to initiate enormous numbers of individuals, such as in Great Britain and Ireland.

Figure 8.6 shows that, for the Dutch population, 14 years after establishment (1950–1963) population growth was still exponential and there was no sign

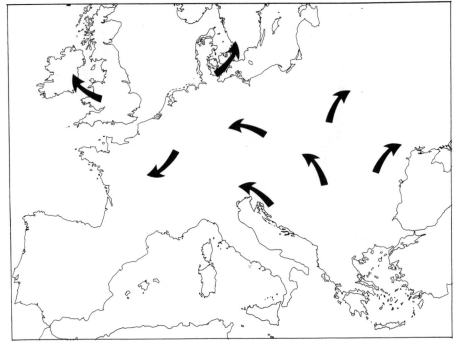

Fig. 8.4 Main routes of spread of the collared dove into Europe (after Stresemann and Nowak, 1958).

Table 8.1 Population increase of the collared dove in Great Britain

Year	Individuals			Locations			
	absolute (n)	*log n*	$\dfrac{log(t+1)_n}{log(t)_n}$	*absolute (l)*	*log l*	$\dfrac{log(t+1)_l}{log(t)_l}$	$\dfrac{log\ n}{log\ l}$
1955	4	0.60		1	0.00		
6	16	1.20	2	2	0.30		4.00
7	45	1.65	1.37	6	0.78	2.58	2.12
8	100	2.00	1.21	15	1.18	1.51	1.69
9	205	2.31	1.16	29	1.46	1.24	1.58
1960	675	2.83	1.22	58	1.76	1.20	1.61
1	1900	3.28	1.16	117	2.07	1.17	1.58
2	4650	3.67	1.12	204	2.31	1.12	1.59
3	10200	4.01	1.09	342	2.53	1.09	1.58
4	18855	4.28	1.07	501	2.70	1.07	1.59
1970	$(95\text{–}158) \times 10^3$	4.98–5.20	1.19				
1972	$(190\text{–}253) \times 10^3$	5.28–5.40	1.05				

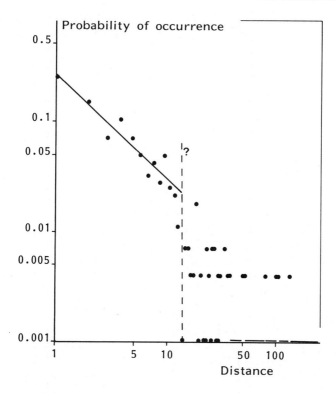

Fig. 8.5 Dispersion probability field of the collared dove as the relationship between the probability of occurrence of a breeding pair at various distances from their parent's nest (after Hengeveld, 1988a).

of levelling off. In Great Britain growth rate was exponential for a decade, and then appeared to decline between 1965 and 1975 (Fig. 8.7 and Table 8.1). Moreover, the rate of increase of the number of recorded locations in Great Britain and Ireland decreased slightly through time. Interestingly, this rate became identical to the growth rate of the number of birds recorded. Also, the ratio of these two rates, expressing the species' spatial saturation, shows the same trend, being high at first and later decreasing to a lower value (Table 8.1). Although all three rates decline, growth remains exponential. The curvilinearity of Fig. 8.7 could be interpreted in terms of logistic growth, as, for example, Hutchinson (1978) suggested, although its descriptive formula should be replaced by a mechanistic one to give insight into what is happening. However, the rate of spatial advance of the wave front cannot be explained by a logistic saturation of the newly attained area, the populations in this region still being in their exponential growth phase.

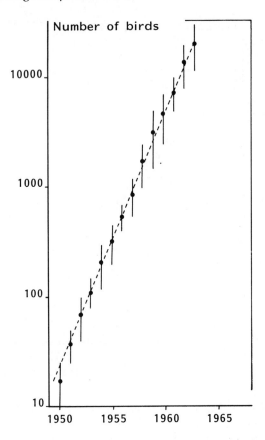

Fig. 8.6 Growth of the Dutch population of the collared dove over the first 13 years after its arrival (after data in Leys, 1967).

8.1.2 Modelling the expansion rate

One of the main problems with Skellam's (1951) model is that the constant population level cannot be estimated independently. In the collared dove population growth cannot be described in terms of the logistic equation. Another concern is that individuals are assumed to keep moving in Brownian fashion throughout their life. More important, though, is the fact that, after settling of the juveniles, the birds are sedentary for the rest of their life.

The stochastic model discussed in Chapter 7 and applied there to the epidemiology of stripe rust, was also applied to the European invasions of the collared dove and the muskrat (Van den Bosch *et al.*, in prep.). Apart from one parameter value for net-reproduction, several parameter values are also calculated for the probability densities of both the species' reproductive

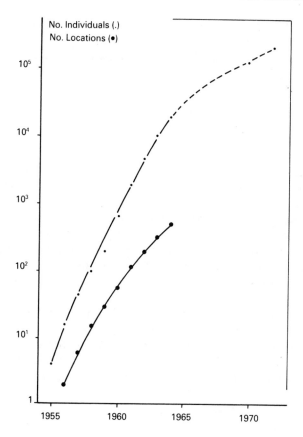

Fig. 8.7 Population growth of the British population of the collared dove over the first 10 (Hudson, 1965) or 15 years (Hudson, 1972) after its arrival. The large dots give the number of locations occupied (after Hengeveld, 1988a).

age and the distance of settling from the parent's nest. These statistics are the mean and variance for reproductive age, and the variance for the spatial dispersion of offspring around their parent's nest (Table 8.2). The estimates of these parameters can, in turn, be derived from information on survival rate, reproductive fraction, number of fledglings per breeding female and the fertility per half-year age class (Table 8.3). The result of these calculations is an estimate of the expected wave velocity. This expected velocity is 56.3 km per year. This value can be compared with the observed one from the regression of the square root of the area occupied against time, which happens to be smaller by 29%, i.e. 43.7 km per year.

Density-dependent factors are of no concern because the populations in the wave front are still in their exponential growth phase where logistic saturation effects are negligible. In the collared dove this phase lasted more

Table 8.2 Number of ring recoveries, in various distance classes from the place of marking, for the collared dove (Streptopelia decaocto)

Class boundary (km)	0	50	100	150	200	250	300	350	400	450	500	550	600	650	total
Number of recoveries	38	8	5	3	6	6	1	0	1	1	0	1	2		72

than 10 years in Britain, and 15 in the Netherlands. Taking this period to be 10 years, and wave velocity 45 km per year, the wave advances at least 450 km before density effects in the original locality come into play. But by then they cannot possibly affect the speed of wave progression at the front, given the much smaller radius of the dispersion probability field.

However, local population increase is not purely exponential; part of it is due to immigration, although the influx of individuals adding to local population increase has so far been underestimated (e.g. Okubo, in press). Quantitatively the net result of both processes is indistinguishable from exponential growth, although qualitatively it differs. Yet, the growth rate of the starling near its time and point of liberation in New York is lower than the later growth rate in surrounding states with an immigration component (Fig. 4.10). In the collared dove the components cannot be separated in this way. Still, as the steepness of the expansion wave determines local popu-

Table 8.3 Life-table statistics for the collared dove (Streptopelia decaocto)

Age interval d (yrs)	Survivorship $L(d)$	Reproductive fraction	Number of fledgelings per breeding female	Fertility $M(\alpha)$	Total reproduction $L(d)M(\alpha)$	$B(\alpha)$
$0-\frac{1}{2}$	0.86	0.0	3.125	0.000	0.000	0.000
$\frac{1}{2}-1$	0.52	0.1	3.125	0.313	0.163	0.061
$1-1\frac{1}{2}$	0.31	1.0	3.125	3.125	0.969	0.363
$1\frac{1}{2}-2$	0.23	1.0	3.125	3.125	0.703	0.264
$2-2\frac{1}{2}$	0.13	1.0	3.125	3.125	0.403	0.151
$2\frac{1}{2}-3$	0.07	1.0	3.125	3.125	0.214	0.080
$3-3\frac{1}{2}$	0.04	1.0	3.125	3.125	0.123	0.046
$3\frac{1}{2}-4$	0.02	1.0	3.125	3.125	0.063	0.023
$4-4\frac{1}{2}$	0.01	1.0	3.125	3.125	0.031	0.012

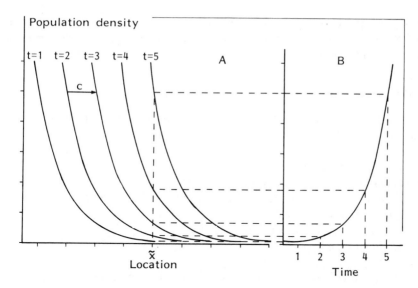

Fig. 8.8 Relationship between the steepness of the invasion wave and local population growth (after Van den Bosch *et al.*, in prep.).

lation increase (Fig. 8.8), estimates of the expected increase can be compared with that of the observed one, as in the case of wave velocity. These values are 0.32 and 0.51 per year, respectively, differing by 37%. In this case the expected rate is smaller than the observed one, whereas the reverse holds for the velocity rate. In order to reduce this discrepancy, improved data or additional parameters should increase the expected values of local population increase and at the same time decrease that of the propagation rate of the wave front.

The differences between the expected and the observed rates of wave velocity and population increase can be caused by one or more of three factors: (1) the yearlings are less vital than older birds, (2) not every adult breeds, and (3) the estimated dispersal density is inaccurate. They all reduce the expected wave velocity, but only the first two increase the rate of population build up, as required. Without new information, however, it is not possible to reconcile the differences between observed and expected rates.

Thus, contrary to Skellam's (1951) model, testing of expected values against observed ones has become feasible, allowing improvements to be made, either by making the ecological data more accurate or the set of explanatory variables more complete. Improvements to the structure of the stochastic model could be (1) the relaxation as to density dependence, and (2) making it an analytical model. This means that, unlike many other ecological models, it is pertinent not only to the individual cases for which it was formulated but that it is generally applicable. This is because the form of the equations used has been explicitly derived from general mathematical theories on wave propagation.

8.1.3 Explanations

Six explanations of the collared dove invasion can be proposed (cf. also Hofstetter, 1960), namely:

1. birds enter open communities
2. resistance of closed communities is overcome
3. effects of natural enemies, predators or diseases are absent or reduced
4. food conditions improve
5. the species' genetic constitution alters
6. the climate improves.

The first four explanations are demographic and assume a constant climate, as does a genetic explanation.

How changes in demographic parameters work is difficult to see, either in the species itself or in the communities penetrated. Resistance of 'closed communities' is difficult to estimate independently from their invadability, resulting in circular reasoning. No data exist to test these explanations. Rucner (1952), for example, suggested that invading Europe was possible because corvids, natural enemies of the collared dove, were absent, but he did not give details on actual effects of their possible interaction. Competition with its closest relative in these regions, the turtle dove, *Streptopelia turtur*, does not occur because the two species live in different habitats. Food conditions may have improved, suggested by the fact that high concentrations are found at the grounds of grain terminals. However, expansion started during the last part of the 19th century and in regions without increasing numbers of terminals. Extension of European farming cannot explain the dove's Middle Eastern expansion either, nor its confinement to human settlements in Western Europe. Stresemann and Nowak (1958) suggested a change in the dove's nesting behaviour regarding its breeding in trees in Europe and presumed a heritable change. This supposes increased tree growth in the Middle East and a genetic basis (Mayr, 1951), which are both unproven.

Glutz von Blotzheim and Bauer (1980) equated this invasion to climatic changes during this century. Warmer conditions, particularly in winter, could be advantageous to this subtropical species. A more equable European climate allows an extensive reproductive period of up to 9–10 months per year (cf. also Hudson, 1965), even longer than in most parts of its original range. This hypothesis is testable as it concerns net-reproductive output under various conditions and in various parts of the range including outside

Europe. Also, recent developments in synoptic climatology have shown great climatic variability (e.g. Lamb, 1977) to which individual species in Europe (e.g. Ford, 1982) or whole floras (Erkamo, 1956) and faunas (e.g. Hengeveld, 1985; Kaisila, 1962) responded during the same decades.

So far none of these explanations has in fact been tested. It is possible that factors related to reproductive output are particularly important bearing in mind that reproduction is one of the parameters in the model explaining most of the rate of invasion, and that net reproduction per female is greater than one ($R = 1.33$). There is no reason to assume that dispersal of yearlings around their parent's nest as the other explanatory process has increased simultaneously throughout the range.

8.2 Four other European invaders

During the same period, four other bird species among many others have invaded Europe: the serin (*Serinus serinus*), the penduline tit (*Remiz pendulinus*), the scarlet rosefinch (*Carpodacus erythrinus*), and the fulmar (*Fulmarus glacialis*).

Mayr (1926) was the first to draw attention to the invasion of the serin, a bird from northwest Africa and southern Europe. After 1800 it spread north of the Alps and in 1925 to Central Europe (Fig. 8.9). First it followed two routes, a western one through the Rhine valley and an eastern one through that of the Danube. Then its expansion stagnated until the late 1950s or early 1960s; in Denmark it even withdrew during the 1940s (Olsson, 1969). After the early 1960s it expanded again, keeping to its initial northeasterly direction. Apart from expanding its range across Europe it also expanded into Turkey, the Caucasus and southwestern Iran. Its eastward expansion led it into Romania and the Ukraine (Olsson, 1971). The whole expansion period covers at least the 200 years with written records (Olsson, 1971).

When expanding, it occupies the most preferred habitats first, colonizing them at a high rate; later it penetrates into less preferred habitats, thus filling up the available space. This results in a broad border zone occupied by a few young individuals or breeding pairs of young birds around the region of more densely populated areas. However, its habitat choice varies over its range; although Mayr (1926) does not mention pine forest on sandy soils as a habitat, the serin prefers this in the Baltic countries, being frequently observed there in southern Sweden (Olsson, 1971). Also, in the northeastern parts of its present range it has become a migrant, whereas in the southern parts it is still sedentary.

More recently, the penduline tit expanded its range westwards into Central Europe (Fig. 8.10) (cf. Flade *et al.*, 1986), keeping to the latitude of the main part of its geographical range. During the 19th century it was still confined to Poland and nearby coastal areas along the southern Baltic. At some point it started to expand westwards which happened in distinct

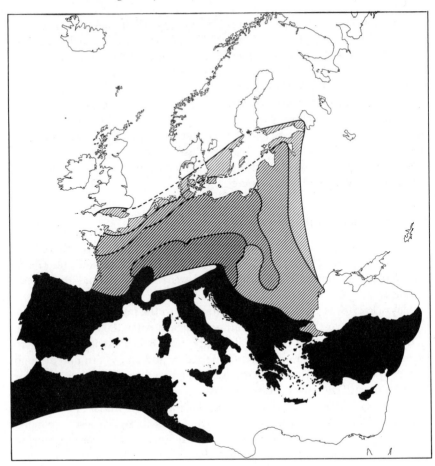

Fig. 8.9 European expansion of the serin up to 1970 (after Olsson, 1969).

temporal waves, rather than continuously. Thus, it almost ceased expanding between 1965 and 1975, after which it spread at a high rate in westward (about 250 km) and northward (about 200 km) directions. Earlier expansion waves occurred about 1935 and from 1950 to the early 1960s. Moreover, not only did it expand discontinuously in time but in space as well. Thus, one such wave during 1978–1980 in Bavaria can hardly be found further north in Germany, in Lower Saxony and Schleswig-Holstein. To the contrary, the stagnation of 1985 in northwestern Germany, following the rapid increase of 1982–1984, coincided with the strongest wave in southern Germany. Also, contrary to the buildup of its Central European range since the mid-1960s, the French populations declined in that period.

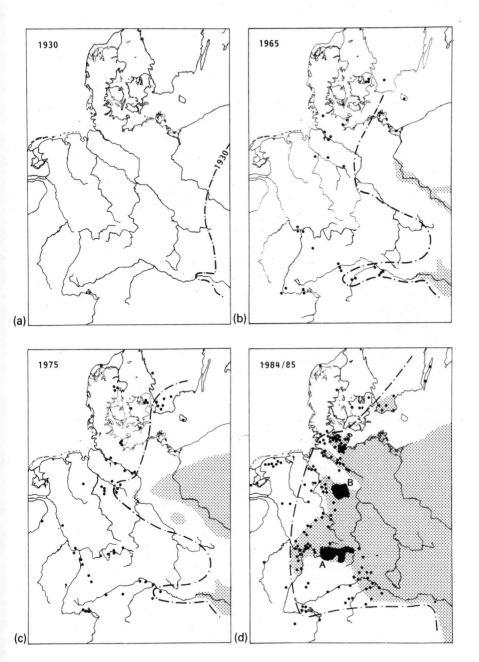

Fig. 8.10 European expansion of the penduline tit. Hatched: closed part of the range (after Flade *et al.*, 1986).

Similar to invasions of other species the more or less closed front was preceded by establishments of single individuals or of isolated breeding pairs (Fig. 8.10). These birds, forming bridgeheads for further progression of the more closed front, followed the east–west valleys of main rivers, such as the Elbe, the Aller, the Main and the Danube. Earlier, the species expanded westwards along the Baltic coast, southwards along the Vistula and later along the Rhine in the west. This expansion results from dispersing birds covering about 40 km or, occasionally, 100 km from their source area, and happens between, as well as during the breeding seasons. Fig. 8.10 shows that, as in the serin, the range is spatially saturated at some distance behind the wave front. Thus, at least in these cases, one can think in terms of wave zones rather than in those of more sharply delimited fronts.

The scarlet rosefinch is another species spreading westwards since the 1930s. In fact, it expanded westwards in the beginning of the nineteenth century, contracted in the 1840s, expanded westwards after the turn of the century and contracted eastwards again (Newton, 1985). Since the 1930s, it spread consistently, first into northern Europe and later into Central Europe (Bozhko, 1980; Stjernberg, 1979, 1985). During the 1930s and 1940s, it increased its numbers and reached high densities during the 1960s and 1970s. Also in the 1930s, it expanded westward along the Baltic coast and into Finland, where it expanded rapidly during the 1940s, becoming widely distributed in the 1950s and 1960s. From there it reached Sweden in the 1950s where it spread widely and built up its numbers in the 1960s and 1970s; in 1970 it reached Norway. In 1966 its first breeding pair was observed in Denmark where it increased during the next decade. Since the 1950s and particularly during the following two decades, it spread south-westwards into Central Europe, where it first became a regular breeding bird in Poland. Since the 1960s it occurs in Czechoslovakia, since the 1970s in Austria, and since 1978 in Yugoslavia. After the late 1970s it expanded into Bulgaria, Switzerland, France, Great Britain and finally Scotland, where it first nested in 1982.

Stjernberg (1979), studying its population dynamics, thought that environmental changes account partly for the spread of the scarlet rosefinch, but that its properties contribute to it taking advantage of these changes. In the first place, the higher temperatures of particularly the 1930s caused the vegetation to develop earlier in the year, becoming more suited to the bird's needs when it returned from its wintering grounds. Apart from this, almost the same period, the 1920s and 1930s, also saw the decline of the linnet (*Carduelis cannabina*), its possible competitor in Finland. Finally, the landscape changed during that period, resulting in a higher frequency of the bushes it prefers as a habitat. Properties that could facilitate its expansion are low adult mortality, strong site tenacity of second-year birds, breeding sociability, wide biotope choice and unspecialised diet, and flock migration particularly of adult females and young males. These females and young males arrive in the newly colonized regions after the adult males, which disperse solitarily. After arrival they often found populations outside the previous breeding range, which later become centres of secondary spread.

Thus, both habitat changes and the species' properties resulted in higher breeding success and this, in turn, in its westward expansion. However, the habitat changes the scarlet rosefinch may have taken advantage of occurred particularly in Finland during and after the 1930s. It is not clear whether they also explain its growth and spatial expansion in Russia and Europe under the cooler conditions since the mid-1950s, and the contractions and expansions after about 1800.

The fulmar is the fourth species that invaded Europe; it expanded from the north over a period of about 200 years. Fisher (1952, 1966) meticulously described its expansion and local population growth that started at islands in the North Atlantic, then along the Icelandic coast, and then, via St. Kilda, to the Faroes, the Hebrides and the Shetlands along the British and Irish coasts.

The first-known record of the fulmar is from Grimsey north of Iceland in 1640. After that time, in 1750, it invaded Iceland, in 1820 the Faroes, in 1878 the Shetland Islands and after this, towards and at the turn of the century, Scotland. Then it gradually spread southwards, particularly along the west coasts of Britain and Ireland and to a lesser extent along the east coast. Since 1921 it occurs along the Norwegian coast (Haftorn, 1971) and at present it even breeds along the south coasts of England and Ireland and since 1960 along the French coast (Cramp *et al.*, 1974). Figure 8.11(a) shows its spread in Britain and Ireland, illustrating the variation in both the length and the direction of the steps taken. Often it leapt forward a great distance and then spread in both directions along the coast, thus filling in the gaps and advancing the range.

After the various local populations were founded they grew exponentially. Figure 8.11(b) not only shows the similarity of these rates but also that even after 90 years the colonies were still growing exponentially, despite their often huge size. Only the growth rate of the first-founded and largest colony in the Shetland Islands is decreasing. The relatively low values for 1959 in several colonies seems not to indicate saturation, but is probably due to misleadingly low counts in that year (Cramp *et al.*, 1974).

Fisher (1952, 1966) explained this expansion by the amount of offal produced, at first by whalers and later by trawlers (see, however Brown 1970). Contrary to this Wynne-Edwards (1962) hypothesized either a change in the species' genetics or in its behaviour. Finally, Salomonson (1965) thought that a warming of the northeastern Atlantic explained its expansion, although Sharrock (1976) thought that this could not hold because some early stages coincide with two cold periods in the 19th century. This may seem worse than it actually is, as the main surge of expansion and population

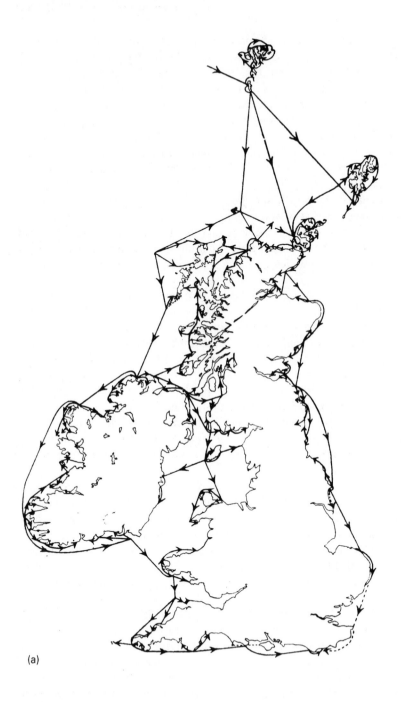

(a)

Fig. 8.11(a) Expansion of the fulmar along the British and Irish coasts (after Fisher, 1952).

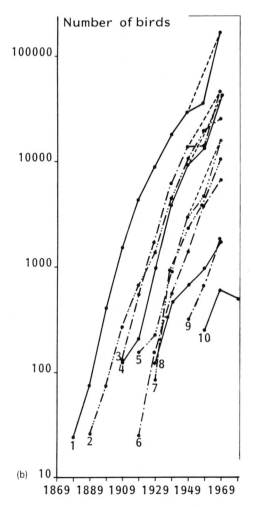

Fig. 8.11 (b) Population growth in various British fulmar populations (after data in Cramp *et al.*, 1974).

growth, including those in Iceland and the Faroes, occurred during the 20th century (Fisher, 1952).

8.3 Two American invaders

Particularly in North America, two bird species are attracting much attention, the cattle egret (*Bubulcus ibis*) and the house finch (*Carpodacus mexicanus*). Of these, the cattle egret is a natural transoceanic invader from Africa, whereas the house finch has been transferred from the American

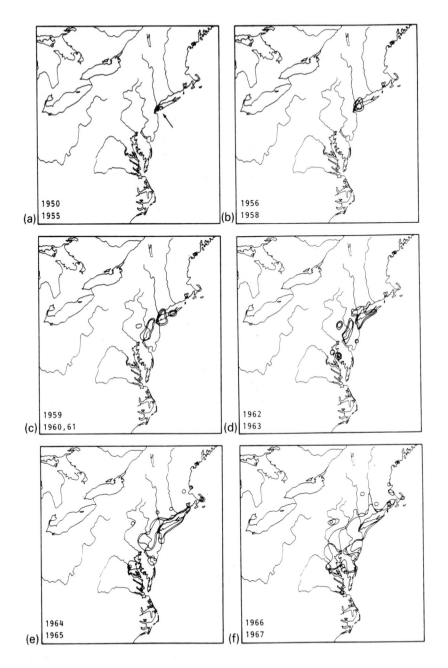

Fig. 8.12 Expansion of the winter range of the house finch in eastern North America, 1949–1979 (after Mundinger and Hope, 1982).

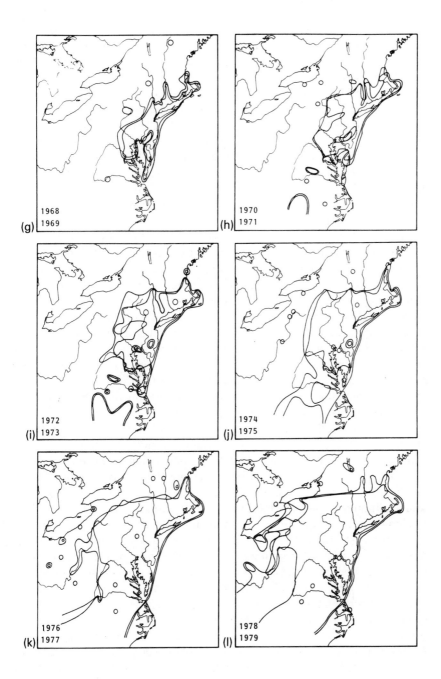

west to the east by man. Spectacular invasions from Europe concern the starling (Figs 4.7–4.10) and the house sparrow (e.g. Dott, 1986; Robbins, 1973; Wing, 1943).

The house finch was introduced in 1940 on Long Island and increased at an accelerating rate in the subsequent years (Mundinger and Hope, 1982). From 1966 to 1979 it increased at a rate of 21% annually (Robbins *et al.*, 1986) and in the 30 years from 1948 to 1978 its range increased about a thousand-fold from 177 miles^2 to 169120 miles^2. As with other invaders, the house finch progressed by both long-distance dispersal and neighbourhood diffusion, following rivers and coastlines. Apart from this, it can also be found in urban environments, causing its spread along the Boston–Washington corridor.

Figure 8.12 shows the progression of the house finch from 1950 to 1979. First the maps show that it spread from focal points in all directions. All maps show these points as isolated bridgeheads ahead of the main front of the expanding winter range; often these bridgeheads are centres of expansion during the next years. As these centres expand they gradually fill in space until they are taken up in the main part of the range.

The cattle egret first expanded in Africa in various directions. For example, during the 19th century it did not occur in South Africa where at present it is common. At the turn of the century it reached the Cape, where it possibly bred from the 1920s onwards. Between 1920 and 1940, and particularly during the 1930s, it built up its population there after which it seems to have been stable (Siegfried, 1965, 1966).

Possibly during the mid-1930s it crossed the Atlantic Ocean to Guyana on the north coast of South America (Handtke and Mauersberger, 1977). Simultaneously with its expansion in South Africa it invaded North America via Florida, penetrating first northwards along the east coast (Fig. 8.13). In South America, to the contrary, it first spread westwards and then to the south, although it seemed incapable of occurring in most of Chile. After reaching northern Argentina it expanded northwards and southwards along the east coast, thus encircling the Amazon Basin. Similarly in North America, it at first left the inner parts of the continent uncolonized, later penetrating them from the east. From the west, its expansion was probably halted by the deserts east of the Rocky Mountains.

All this information pertains to the subspecies *Bubulcus ibis ibis*, but a similar expansion occurred during the same period in the more eastern subspecies *B. i. coromandus* from its range in India and the Far East into Australia and New Zealand (Handtke and Mauersberger, 1977). In Australia too, it expanded in a broad, coastal belt, leaving the interior of the continent unoccupied.

For several reasons, Handtke and Mauersberger (1977) feel that the cattle egret seems well-suited for this rapid range expansion. It can reproduce

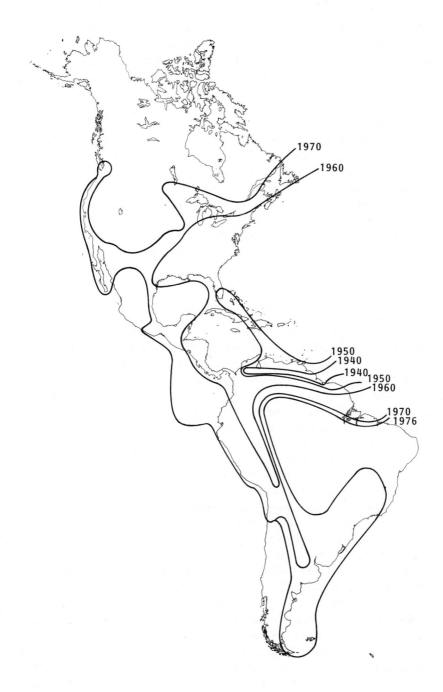

Fig. 8.13 Expansion of the cattle egret in North and South America (after Handtke and Mauersberger, 1977).

quickly and disperse over long distances and it can occur close to human settlements, being found in cattle-breeding areas. It breeds in mixed colonies with other heron species, which may decrease its mortality rate even from the first stages of settlement onwards. Breeding is apparently initiated by rainfall, thus allowing the species to breed twice a year and also during periods when other species do not breed. Thus, it suffers little from inter-specific competition. Also, it seems not to compete for nesting sites, nor intraspecifically for food. Predation as a biotic factor seems to be the only important cause of mortality for this species. Abiotic factors such as drought and cool conditions can limit the cattle egret.

Thus in these two invaders, man is involved; in the house finch as a transfer agent, and in the cattle egret as a factor facilitating its living conditions. The first species exhibits processes at an intermediate spatial scale, showing up spatial saturation after long-distance dispersal, whereas the second species shows range expansion at an extremely broad spatial scale.

8.4 Conclusions

An interesting point is that the expansion directions vary greatly in these species, although they expanded partly or wholly during the same period. In Europe, the collared dove expanded northwestwards, the serin north-eastwards, the penduline tit and the scarlet rosefinch westwards, and the fulmar southwards, these directions being perpendicular or even running opposite to each other.

Secondly, long-term studies in different regions within the newly acquired parts of the range of both the collared dove and the fulmar show that the exponential growth phase can last rather a long time. This implies that density-dependent factors cannot account for the expansion. Also, neither community resistance to invaders, nor genetic or behavioural changes underlying a species' ecology have so far been shown conclusively to account for the expansion of any species. The invasion front formed by these early settlers is only closed on a broad spatial scale; at finer scales it shows up isolated bridgeheads at various distances ahead of a more closed breeding area. The often-observed widened habitat choice some time after the arrival of the first invaders could be due either to imprinting of new habitat conditions or overflow from preferred, optimal habitats into more marginal ones. These might not explain range expansion, but rather be an exhibition of its side effects. Habitat choice often increases from the range margin towards the centre, not only in animals but also in plants (Hengeveld and Haeck, 1981) to which neither the imprinting theory, nor the overflow theory apply. Rather than behavioural or demographical factors, physiological responses to climatic factors, however they operate, seem to be more likely.

Thirdly, Stjernberg (1979) observed that females of the scarlet rosefinch are more mobile than males as in other bird species such as the reed bunting (*Emberiza schoeniculus*), the blackbird (*Turdus merula*) and the pied flycatcher (*Fidecula hypoleuca*). This seems to be general in birds, and different from mammals where males are the more active sex (cf. Table 8.1 in Shields, 1983). Two examples of this, referred to above, are the red deer in New Zealand and the muskrat in Europe (cf. Lubina and Levin, 1988, for the California sea otter).

Finally, the five bird species selected are only a sample of a larger number of species showing geographical range dynamics (e.g. Järvinen and Ulfstrand, 1980; Niethammer, 1951; Von Haartman, 1973). For example, in Europe the common grebe (*Podiceps cristatus*) recently expanded northwards; the cettis warbler (*Cettia cetti*), the sand martin (*Riparia riparia*) and the starling (*Sturnus vulgaris*) contracted southwards; and during the last century the citrine wagtail (*Monticilla citronella*) expanded westwards in Siberia for a distance of 3000 km. Furthermore, the kittiwake (*Rissa tridactyla*) expanded southwards and now seems to be contracting northwards again. North American species show similar geographical dynamics, with the best-known example being the cattle egret (*Bubulcus ibis*) in America (Handtke and Mauersberger, 1977). Also, the house finch (*Carpodacus mexicanus*) expanded rapidly in the eastern United States after its introduction there in 1940 (Mundinger and Hope, 1982), although it declines in its original range in the west (Robbins, 1988). Changes like these are also known for other taxa, such as ground beetles in Europe (Hengeveld, 1985; cf. also Ford, 1982 for examples of many individual species).

All these changes in abundance and distribution show up species responses to ever-changing environmental conditions. Invasions are not exceptions, but the rule. They are not isolated and incidental phenomena disturbing the order of ecological communities; they exhibit the dynamics inherent to any species, be it on a broad or a fine spatio-temporal scale. Moreover, the dynamics will to a large extent be stochastic, as simulated in that of the wave progression in the collared dove.

Nine

The stochastic structure of the wave front of rabies in central Europe

Often it is difficult, if not impossible, to tell where the front of a species' wave of advance runs. If some front can be recognized it should be defined, as in the penduline tit or the house finch, in terms of a broad zone of scattered populations surrounding areas of more closed occurrence. But also in these two species breeding pairs or isolated individuals settle far ahead of this vaguely defined zone. These isolated settlements are bridgeheads from where secondary, neighbourhood dispersal starts, making the spreading process stratified.

Diffusion is a stochastic process defined by spatial transition probabilities that are estimated from the frequency with which various distances are covered. Thus, from the number of new breeding pairs or plants at particular distances from the parent's nest or population one can derive the transition probability for those distances. All probabilities together for a range of distances define what I have called the dispersion probability field around the parent's nest or population (Chapter 5). In other contexts this concept is known as contact distribution, primary gradient, dispersal density, spatial expectation density function or information field, but these terms are either inappropriate or too vague and clumsy. Similarly, ecological parameters, such as reproduction or mortality can be defined statistically, giving, respectively, reproduction and survival probabilities for various age classes. In short, ecological parameters relevant to the diffusion process can be defined in statistical terms and hence the process of biological spreading as a whole as a stochastic process.

Because of this one expects to find transition zones rather than sharp fronts of distinctly advancing waves. One also expects their progression to be jerky rather than steady depending on topographical and climatological factors and their interactions. Occasionally, one can analyse the stochastic structure and advance of a wave front in enough detail which is the case in the advance of rabies epizootic in central European foxes. Two approaches are feasible, one in global terms describing the front as a uniform wave, and the other in focal terms describing it as the progression of several individual foci that keep their individuality whilst moving. These foci, moreover, are structured in time.

9.1 The progression of the wave front

Sayers *et al.* (1977; cf. also Sayers *et al.*, 1985) mapped approximately 3000 confirmed cases of rabies collected in central Europe in the years 1963–70. They recorded them in a 133 × 133 km² lattice subdivided into 32 × 32 square grids, each containing information on monthly rabies incidences. After a preliminary analysis including data smoothing, they estimated the dispersion probability field from 1219 cases in three-monthly periods during the first quarter of 1965 to the last quarter of 1966 (Fig. 9.1).

Thus, they constructed iso-probability maps describing the probability of new incidences at various distances from the existing ones (Fig. 9.2). Next, they drew the iso-lines of certain selected probability levels, those of *P*=0.55, *P*=0.65, and *P*=0.80, giving an array from very broad to rather fine-scale pictures of spatial progression, respectively (Figs 9.3(a), (b) and (c)). The broadest scale of *P*=0.55 shows quite well-defined, regular waves of

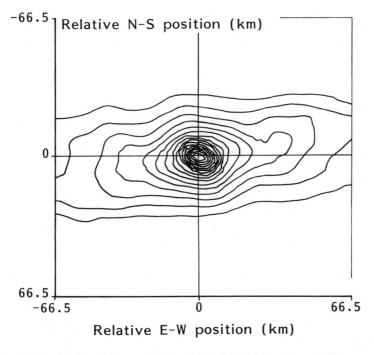

Fig. 9.1 Estimated dispersion probability field of rabies based on 1219 case occurrences during 1965–1966 in southern Germany. The relatively steep north-south decline indicates a greater velocity than from east to west (after Sayers *et al.*, 1977).

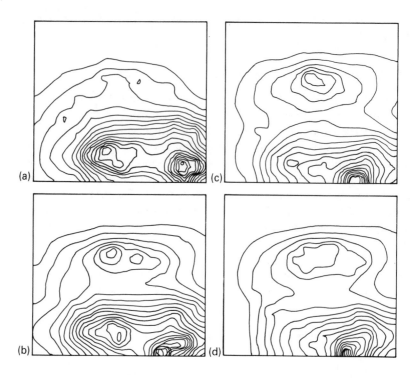

Fig. **9.2** Estimated densities of case occurrences of rabies for four successive 3-month periods. Maximum density is 100%; density contours are given for 5% intervals. The lowest contours indicate the global wave progression; the local, irregular high densities indicate individual disease foci (after Sayers *et al.*, 1977).

advance, whereas the finest scale of $P=0.80$ shows isolated pockets of rabies incidence which shift from one period to the next. Thus, on the broadest scale, the wave of advance appears as a closed, uniform front, but this front breaks up when viewed at finer scales. However, finer scales are not necessarily the better ones; they just show different aspects of the spatial disease progression. For example, what the finest scale does not show, although it is quite clear on maps of the two broader scales, is the gradualness of the wave progression during periods 8–12, and that of periods 14–18, enclosing an intermediate transition period, 13. On the finest scale these periods appear as distinct spatial clusters supplemented by two other clusters, that of periods 14 and 15 at the central right, and that of period 20–23 in the lower right-hand corner. Interestingly, the Danube separates the two temporal clusters in space, those before and after period 13. Apparently, this river was crossed during period 13 at two points at the right of the centre of the map.

Thus, Figs 9.2 and 9.3 suggest that this wave front was not uniform at a fine scale of resolution, but consisted of several foci. Moreover,

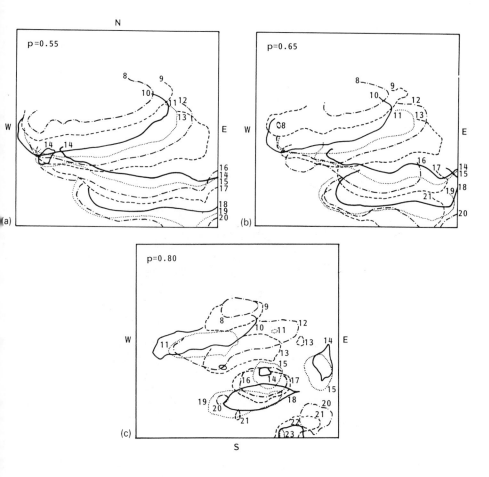

Fig. 9.3 Iso-probability contours of rabies incidence for 13 quarterly epochs running from the autumn 1964–1967. The probability levels in the three figures are $P=0.55$, 0.65, and 0.80, respectively, and range from a more global view to a more focal one (after Sayers *et al.*, 1977).

Figs 9.3 (a), (b), and (c) indicate that the progression rate was not constant over the period, but that it varied.

9.2 Foci within the wave front

Thus, rabies progresses as foci of disease incidence that propagate separately and independently. To analyse this in even greater detail, Sayers *et al.* (1977) applied two-dimensional Fourier analysis, which describes the average spatial dependence of the probability of incidences happening between subsequent periods. This gives new contour maps, this time for intensity

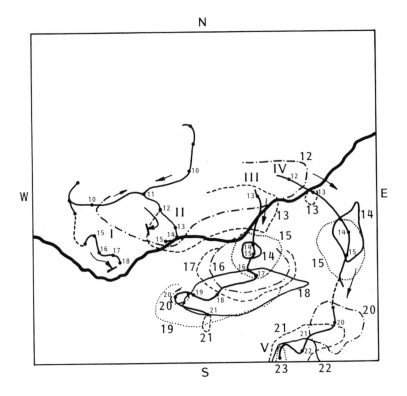

Fig. 9.4 Trajectories of two individual foci superimposed on a map of iso-probability contours of rabies incidence at the level of $P=0.80$, together with the course of the river Danube (after Sayers, *et al.*, 1977).

probabilities of the Fourier function. For periods with hardly any spread, this function has maxima close to each other. If, however, the disease temporally spreads a longer distance, a new maximum appears away from the previous one. Lines connecting subsequent maxima express the spatial trajectory the disease focus has followed. Figure 9.4 shows the trajectories for several foci, together with iso-probability lines for $P=0.80$ from Fig. 9.3(c). This map confirms that during period 13 the Danube was crossed at two points from where rabies spread southwards along two independent trajectories. However, trajectories more to the left on the map stopped at the river.

Thus, foci within the wave front propagate both separately and independently, causing this front to be structured. Moreover, similar trajectories occur in the same region some distance behind the front (Fig. 9.5), indicating that processes happening at the front are also found within a range. Apparently range expansion is not brought about by particular processes; only their result differs. Processes at the margin appear to us as directional

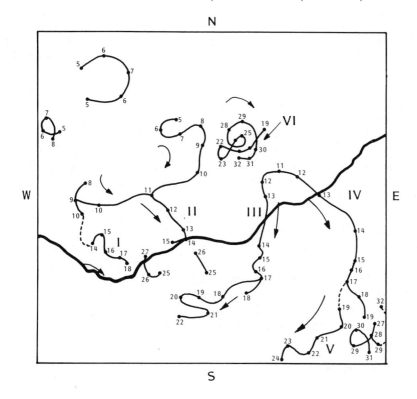

Fig. 9.5 Trajectories of several individual disease foci of rabies in southern Germany (after Sayers *et al.*, 1977).

because, as a result of differences in density, susceptibles move at higher frequencies outwards from the range than towards it.

9.3 Temporal structure of the wave front

After this spatial analysis of rabies progression, Infantosi (1986) analysed more or less the same data for their temporal structure. His time series ran from 1st January 1963 to 31st March 1971 and included 2326 incidences, grouped into 99 months. The area covered a surface of 147 × 133 km². Four steps can be recognized in this analysis:

1. assessment of the temporal structure of the epidemic wave into a linear sequence of independent events
2. analysis of the individual events
3. ecological interpretation of events and impact assessment of the disease
4. reconstruction of spatial spread.

First, Infantosi partitioned this area into subunits to make the individual epizootic events independent. Overlap ceased to occur at a subdivision of the area into 64 squares of 305 km², thus containing isolated epizootics. These isolated epizootics were aligned relative to their peak values and then averaged because of their differences, giving average epizootics. These, in turn, were standardized for various time segments to remove effects of non-stationarity of the series, giving rise to coherent average epizootics. Grouping the information into four geographical categories shows that these differ in magnitude of rabies incidence. Also, at this level, a central peak appears to be accompanied by two subsidiary peaks at a distance of ± 12 months.

Thus, rabies epidemics consist of a linear combination of basic structural wave forms of different intensity and with latencies of varying length between them. The intensities could relate to the ecology of the regions and the latencies to the biology and dispersal of the foxes. Infantosi called these basic wave forms microepizootic profiles. The next step he took was to analyse this structured unit even further using a signal analysis technique, inverse filtering. Microepizootics in time series are then transformed into a sequence of impulses at particular locations and with weights equal to their magnitude. Again the resulting microepizootics differ from each other. However, as expected, their features relate to various phases in the fox's life history.

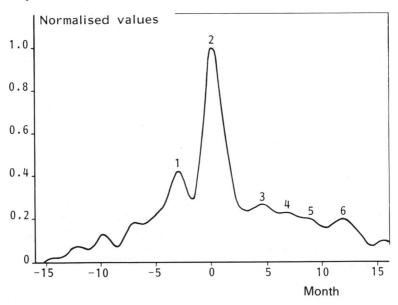

Fig. 9.6 Average course of the microepizootic profile of rabies derived from 78 record segments extracted from 64 subregional records in southern Germany. The individual peaks coincide with particular stages of fox reproduction during the year (after Infantosi, 1986).

He concluded that over the year the profile consists of two main peaks plus several minor ones (Fig. 9.6) that can be related to dispersal, and rutting and mating, respectively. The smaller, first peak precedes the larger one by about three months, relating it to previous dispersal in November and December. The smaller peaks after the main ones relate to other stages in the life cycle with a time-lag of one month of the incubation period of the disease. Thus, the first minor peak in July relates to the end of denning of the young in June when they start to be found within 1 km from their parent's territory. Similarly, the second of these peaks in September relates to pup rearing in August, and the peak in November to the end of this stage in October. These latter phases relate to extensions of the range of the pups to 6 km and 10 km from the parent's territory, respectively.

The microepizootic profile can therefore be interpreted in terms of changing contact rates between foxes, complicated by a time-lag imposed by the incubation period of the virus. The various contact rates, in turn, relate to different parts in the fox's life cycle.

Differences in these microepizootic profiles of the record segments can be averaged out, and the contribution of some background cyclicity removed. The resulting curve can be used in modelling various features of rabies in foxes in southern Germany. It then appears that 83% of the introductions of rabies in new areas occur in the 3rd and 4th quarter of the year, derived from the fox's life history and the virus' incubation period. The greatest magnitude of the epizootic follows dispersal of foxes intruding into un-affected territories and relates positively to population density in healthy fox populations. Following its introduction, rabies reduces the fox population by about 50%, after which the relationship with population density ceases. The intervals between microepizootics are independent two years after the introduction of rabies into a fox population and are only weakly dependent over periods of four years after introduction. Because of reduction in population density of the foxes, their contact rate decreases, leading to a decrease in rabies incidence. Eventually the disease can disappear from the region.

Finally, using the microepizootic profile, the spatial progression of rabies can be reconstructed. Here the first occurrence of the smaller peak was taken to indicate the introduction of the disease. The local presence of a focus was assumed when rabies occurs for at least one life cycle of the fox. Infantosi connected these foci by trajectories according to the local road system, as foxes often follow roads when dispersing. Figure 9.7 shows the resulting reconstruction of the individual rabies foci as based on micro-epizootic profiles which can be interpreted biologically. This figure also shows the general progression of the wave front per fox generation. The rate of progression is around 30 km per year, which is roughly the same as that found independently by mark-release experiments in Denmark (Jensen, 1973).

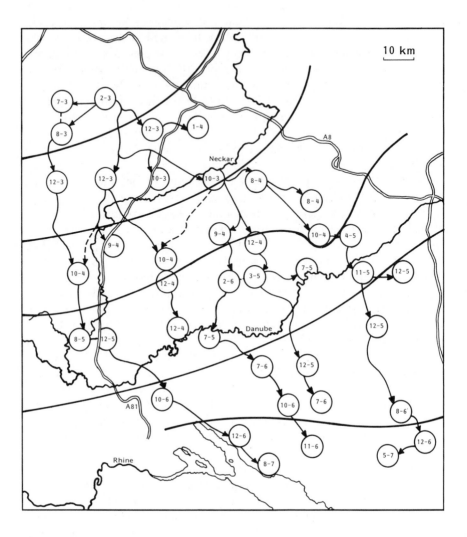

Fig. 9.7 Propagation of rabies in southern Germany during the period 1963–1967 on a monthly basis. The first figure in the circles refers to the month, the second figure to the year within the 1960s. The arrows indicate the roads along which the foxes may have dispersed; the concentric lines separate yearly reproduction periods of the foxes (after Infantosi, 1986).

9.4 Conclusions

The progression of invasion waves is not only structured in space, consisting of foci, but it can be temporally structured as well. Temporal structures can consist of basic building blocks such as microepizootics in rabies forming a complex linear combination. These building blocks, in turn, can also be structured depending on life-history and dispersal characteristics of the species concerned. These again relate to the ecological conditions of the region in which the species lives.

Together the spatial and temporal structures constitute the invasion and determine its characteristics, such as its velocity and the buildup of new local populations. However complex they are, invasions show up a species' spatial dynamics that is even more complicated within its range than when the species is extending outwards.

Ten
Interpreting biological invasions

It is appropriate to interpret invasions in terms of ecological and bio-geographical theories. First, I set the scene by considering the existing ecological paradigm, that of the balance of nature with a population dynamic equilibrium theory as its most important component. This theory is confronted with results of non-equilibrium approaches, among which invasions form an important aspect. Secondly, I consider the historical paradigm of biogeography. Similar to the ecological one, I describe invasions as processes anomalous to this paradigm. After this I discuss recent developments that could eventually lead to a new paradigm for both disciplines. A brief discussion of the implications of this change in paradigm, as to control and conservation measures to be taken, closes the chapter.

10.1 The balance-of-nature paradigm

For a long time a set of theories has dominated ecological theorizing based on the idea that species mutually keep their numbers within bounds. Thus, by evolving or adopting certain functional relationships, structural units result that maximize the species' survival chances and minimize the variance of their temporal fluctuations. These structural units, communities and ecosystems would each be characterized by a particular composition of species that are qualitatively and quantitatively attuned to each other.

Despite its simplicity, this idea is difficult to trace (cf. Egerton, 1973), both in the history of ecology and its various sub-disciplines and in its concepts, definitions and models. Moreover, various criteria can be used to corroborate or to falsify the idea (Table 10.1 and Hengeveld, 1986) resulting in a network of theories, all concerned with or derived from the same idea. As such it is a typical 'paradigm', implying a pattern of theories with a common denominator (Kuhn, 1962), in this case a supposed balance of nature. Through the possible role competition plays in many of the related population biological disciplines, such as evolution theory and population genetics (e.g. Hedrick, 1984; see, however, Harvey and Ralls, 1985 for their discussion of 'evolutionary stable strategies'), also relates to this paradigm with its climax in Mayr's (1982) new species definition, containing niche occupation as a criterion (cf. Hengeveld, 1988a).

Although alternative criteria are put in conjunction in Table 10.1, they have not necessarily been put forward simultaneously. Often, they have arisen in different times and their relative weights have varied among ecological sub-disciplines and individual ecologists. At present, attempts to test this paradigm have intensified, partly increasing the number of criteria and partly resulting in straightforward attacks. Yet most ecologists retain the paradigm.

Because of its heterogeneity, it is difficult to tell to what phenomena this paradigm applies, and anomalies are, consequently, difficult to indicate. However, two characteristics of most of the criteria favouring the paradigm are that (1) demographic factors internal to a system account for (2) its stationarity. Thus, if in disagreement with one or both of these characteristics, invasions are anomalous to the balance-of-nature paradigm, and this has repercussions as to its validity.

10.2 Invaders and the demography of species and communities

Elton (1958), an exponent of the balance-of-nature paradigm, emphasized four ecological parameters as determinants of the success of invaders: (1) a strong resistance exhibited by (2) competing species that (3) structure the community by (4) forming food chains. Thus, demographic interactive processes, particularly competition, are structuring factors.

However, there are few data to substantiate this assertion. Ecke (1954), for example, gives maps that suggest interaction between two rats in south-west Georgia, *Rattus rattus* as the longer-established species and the Norwegian rat, *Rattus norvegicus*, as the more recent invader. These species might compete, although as in Europe, in North America they can co-exist in different habitats (Ecke, 1954). Another invasion possibly involving replacement by interspecific competition concerns the grey squirrel, *Sciurus carolinensis* in England. Simultaneous with its spread the indigenous squirrel, *Sciurus vulgaris*, declined, suggesting interspecific interactions. However, Reynolds (1985) did not find effects of such interactions in 22 annual distribution maps of East Anglia. More recent information suggests that, if interaction does occur, grey squirrels prevent red squirrels from mating, thus reducing their reproductive output. But as in the rats, density limitation as such is not the same as maintaining equilibrium.

Having considered hundreds of species, Lindroth (1957) concluded that only a minority of European insects have settled in North America, and that of these, only a few have spread from their liberation point (cf. Lawton and Brown, 1986). In these species processes of demographic interactions are unlikely to have been restrictive. Mayr (1965) concluded the same for birds,

Table 10.1 Criteria for or against the balance of nature view existing in theories and models in the ecological literature

Balance of Nature	
Proponents	*Opponents*
homeostasis	temporal incoherence
mean density: objective	not real; trends
correction deviation: specific factors	non-specific factors
communities in equilibrium	in disequilibrium
species composition qualitatively, quantitatively stable	not stable
species identical in models	species specific
community homeostatically closed	open
community open or closed: resistance	open, no resistance
community spatially uniform	spatially heterogeneous
community spatially discrete	continuous
community integrated (microcosm, crystal packing, holism, organism)	coincidental
community (dis)harmonic	mixed
community ancient integration	young integration, per (inter)glacial
division of labour	individual struggle for existence
niches (guilds) as functional place	no niches or guilds
selection for co-evolution	against co-evolution
succession: internal community process	forced externally
species, character/size displacement	co-existence
competition	no, or diffuse competition
relationship reproduction/carrying capacity	no relationship; sufficient food
food pyramid	no pyramid; one species in more levels
food chain	no food chain
demographical factors important	climatic or habitat factors important
density dependent factors	density independent factors (liberal regulation)
no time-lags	large effect time-lags
local/laboratory/modelling studies	spatial/geographical studies
one population	several populations
one spatio-temporal scale (competition, density vague)	several scales
spatial dynamics of species coherent	individualistic
processes deterministic	stochastic
natural prevention of outbreaks (when occurring: human causation)	outbreaks normal, regularly occurring
outbreak prevention: introduced species	introduced species fail

(continued)

natural prevention extinctions (when occurring: human causation)	extinction normal
species/characters of colonists and invaders specific	non-specific
viewpoint spatially static	spatially dynamic
hypothesis generation: classification	ordination
general statements (patterns, laws)	individual statements, species unique
deduction from theory	induction from observations
verification	falsification
essentialism	realism

De Bach (1965) and Simberloff (1981a, 1986) for insects purposely introduced for controlling outbreaks of insect pests, and Crawley (1986) for insects and plants. Hengeveld (1986) summarized these studies: 'All communities appear to be open to invasion to a certain extent, although those originated by human disturbance contain most invaders. Literature reviews of several hundreds of species each show that interspecific competition is a dominant factor in a minority of cases only and that supposed invaders are not universally characterized by particular properties. Supposed rigid community structures cannot be shown to exist.' This can be repeated for the studies discussed in the previous chapters.

Similarly, the examples given in the previous chapters do not indicate a dominant role of demographic variables preventing species from invading new areas. Rather, as far as we know at present, species move about freely on all geographical scales.

10.3 The stationarity of species and community processes

The principal aim of present-day population dynamics is to explain why populations neither become extinct, nor grow to outbreak proportions. Instead, they fluctuate within not-too-wide bounds. Thus, time-series are assumed to be stationary, and the remaining task is to explain why the variance of the fluctuation pattern is small. This is the equilibrium theory of population dynamics. Two sub-theories exist within this theory; population regulation and risk spreading. According to regulation theory, through demographic processes with intensities that depend on the size of the deviation from the mean fluctuation level, population density reverts back to this mean level. Alternatively, according to the risk-spreading theory, the variance of fluctuation is kept within bounds purely statistically: the greater the number of mutually dependent variables affecting a species, the smaller the variance. In risk spreading, unlike regulation theory, relationships between species are thought to be non-specific and communities, consequently, to be non-existent. Both theories, as equilibrium theories within the balance-of-nature paradigm, assume spatio-temporal stationarity, meaning that there are no trends, neither in space nor in time (cf. Hengeveld, 1989).

However, Davis (1986) emphasized that on the time-scale of the last two million years of the Pleistocene, temporal temperature trends are found on all scales, up to a period of 100000 years. Yet even the Pleistocene is not representative for a longer geological period, being a cold epoch of a longer-term trend towards global cooling (e.g. Frakes, 1979). Trends in expected density also occur from local to geographical scales, resulting in two-dimensional optimum response surfaces at each of them (Hengeveld and Haeck, 1981, 1982). The two population-dynamic theories should thus be replaced by one assuming non-equilibrium processes, both for explaining individual species processes and for that of community structure and composition (Hengeveld, 1989a, b).

10.4 A non-equilibrium approach

Parry (1978) and Parry and Carter (1985) described temporal variation in oats using economic risk theory, which does not assume equilibrium states to which deviations revert back with minimum variance. Rather, it considers each intensity of the process individually, resulting from the operation of particular causative factors. Thus, oat yield at various altitudes on a Scottish mountain can be expressed in terms of the heat sum of all temperatures above a physiological minimum. In this view, species operate as energy conversion systems, each with specific conversion efficiency parameters. In this conception, the determination of population numbers is not just a demographic but also a physiological process. Local and geographical distribution patterns are viewed as response surfaces relative to intensity gradients of one or more causative variables.

These response surfaces, however, can be conceived in dynamic terms and be treated statistically. Together with temporal variation, weather elements vary in space and so do the species' response intensities. Thus, complicated, temporally shifting polynomial response surfaces can result which, summed over time, assume a more regular shape with highest central expected intensities that taper towards the range margins. Trends in weather conditions result in contractions, expansions or in topographical shifts of these frequency distributions of the intensity of species performance. Given a certain steepness of the peak(s) of the response surface and given a certain change in conditions, local intensities of occurrence either fluctuate hardly or widely.

Central to this non-equilibrium approach is that species can respond quickly in one way or another. Here, we are not concerned with responses in physiology or growth, etc., but more with spatial responses, as these relate to invasions. To match the spatial dynamics of climatic variation, species have to be equally dynamic for finding suitable locations and maintaining populations there, and for recolonizing locations where populations have previously died out (e.g. Pickett and White, 1985). A 'continuous movement of particles' is the mechanism of this dynamic spatial process.

10.5 Invasions as anomalies of the balance-of-nature paradigm

Invasions can be understood within the framework of non-equilibrium approaches, exhibiting continuous spatial adaptations of species to the variability of their environment. But invasions are out of place in the balance-of-nature paradigm. For the conception of communities as sets of functionally co-adapted species assumes that they resist invaders to a certain degree. Community resistance results in inertia to adopt new species or to lose established ones. Moreover, species as integral parts might even be unable to survive without other species of the community. This inertia thus reduces free spatial interchange. Consequently, population dynamic models hardly contain migration parameters, nor leave room for invaders. Neither do they account for species moving about in space, independent of other species, and if they do, as in MacArthur and Wilson's (1967) island theory, the invaders would soon meet the resistance of already established immigrants, through competition. When they managed to establish themselves, their populations would grow out of all proportion before becoming integrated, thus disrupting the existing community for some time (Elton, 1958).

This view is the logical result of a paradigm that emphasises a balance of nature brought about by demographic species interactions. Of course, assuming that effects of demographic factors prevail over those of external factors does not imply that proponents of the balance-of-nature paradigm do not recognize the importance of external factors. They only assume that this proportion is constant, and is thus unable to disrupt a system of species combinations brought about and maintained by mutual interactions.

But when environmental conditions fluctuate or when they show trends, species combinations are easily disrupted. Species responses to external factors are specific, which lessens the possibility of responding immediately and adequately to each other. Rather, the frequency of this happening can be so large that it will select against the formation of functional species relations (Davis, 1976)! As a palynologist Davis is well aware of environmental variability on various time scales. In fact she was one of the first to substantiate non-equilibrium ideas by giving maps of the post-glacial shifts in range location of several tree species (Fig. 2.1; cf. Davis, 1981; Firbas, 1949). As these shifts are specific regarding their origin, direction and rate, invasions inescapably become anomalies to the balance-of-nature paradigm.

10.6 The historical paradigm of biogeography

Up to the present, the idea of spatial stationarity at time scales at least of the period since the last glacial, i.e. about 12000 years, still dominates large parts of biogeography. Again several criteria constitute this paradigm, the most important of which stress an absence or insignificance of dispersal, dispersal barriers, migration routes, allopatric speciation, endemism and, similar to ecological communities, the existence of biogeographical provinces as historical and ecological units. Factors restricting dispersal and the resulting floral and faunal interchange are emphasized relative to facilitating factors.

The idea of spatial rigidity of species stems from early 19th century biogeography when, as first steps in data reduction, biotas were classified into a hierarchical system of spatial units (henceforth called 'provinces'). By then little or no information existed about spatial dynamics of species. Consequently, emphasis was given to borderlines between provinces which, as information accumulated, resulted in lines specific for the various taxa studied (e.g. Holloway and Jardine, 1968; Simpson, 1977). Also, specific explanations were given to these lines, ranging from historical ones and those of restricted dispersal to those of ecological resistance of the province members to invaders from other provinces. Other approaches, emphasizing individual species, attempt to explain endemism (e.g. De Lattin, 1957). Their representatives also emphasize allopatric speciation of, for example, birds since the last glaciation, or genetic depauperation (e.g. Hultén, 1937 for plants, or Lindroth, 1949 for winglessness in carabid beetles). Thus, both approaches within the historical paradigm assume that either barriers prevent regional exchange or that dispersal capacities are insignificant, if present (e.g. Rosen, 1978; cf. also Nelson and Rosen, 1981).

During the last decades several attempts were made stressing dispersal (e.g. Udvardy, 1969). MacArthur and Wilson (1967) were the most significant proponents describing the number of species present on islands in terms of island size, distance to the mainland and time since initial colonization. Time enters the discussion at two levels, (1) that of increasing competition intensity among residents, which would increase with number of new immigrant species, and (2) that of species mutually adapting to each other. Thus, here the condition of restricted dispersal is relaxed, although at the same time it assumes the operation of restrictive processes of the ecological balance-of-nature paradigm. Later, MacArthur (1972) suggested that these ecological restrictions can be so tight that they can explain biogeographical provinces as natural units in terms of resistance to invaders (cf. also Briggs, 1974; Valentine, 1968). If interchange would occur at broader temporal scales, distribution patterns can still be static over sub-periods through a process known as the taxon cycle (e.g. Erwin, 1981, 1985; Howden, 1985).

Thus, even if spatially dynamic processes occur, the end result is still thought to be static. Just as in the ecological balance-of-nature paradigm, species invasions challenge this static view in biogeography.

10.7 Invasions as anomalies to the historical paradigm

As environments are variable, species invasions often occur; one should only be prepared to look for them in order to see them (e.g. Cushing, 1982 for marine fishes; Erkamo, 1956 for plants; Ford, 1982 for animals and plants; Hengeveld, 1985 for ground beetles; Kaisila, 1962 for butterflies; Robbins *et al.*, 1986 for birds). But usually invasions are treated qualitatively by describing a species' occurrence in regions where it was absent before. This qualitative description can be quantified by including geographical density shifts. Thus, from a faunistic one, the approach becomes ecological, epidemiological or geographical, depending on the spatial scale of investigation. For example, Hengeveld (1985) investigating the dynamics of the Dutch ground beetle fauna during the 20th century found considerable changes in the national species composition, paralleling those in climate. During the relatively warm 1930s, 1940s and early 1950s, for example, xerophilous species from southern Europe increased, whereas hydrophilous ones from northern Europe declined. The early 1970s resembled the 1950s more than did the cool 1960s, both faunistically and climatically. Apart from this relatively short-term temperature variation, precipitation gradually increased throughout the century, resulting in a decrease of xerophilous species from early this century onwards and in an increase in mesophilous species during the later decades. This shift is reflected in that of faunistic elements, the southern European elements predominating during the earlier part of the century and the northern ones to the later part. Over this period this fauna was in a constant state of flux, the abundances of the species changing relative to each other all the time.

Thus, quantitatively the faunistic composition altered drastically despite the fact that the qualitative concept of invasion does not apply. Moreover, the static, local view of faunistic and ecological processes does not apply either, which should at least be supplemented by a geographical view defined over a long period of time. As in the ground beetles, in many other taxa such extensions will result in observations anomalous to the prevalent biogeographical paradigm, which should either be supplemented or replaced by a more dynamic and less historically inclined alternative.

10.8 Developments towards a dynamic paradigm

Common to the ecological and biogeographical paradigms is that neither of them regards the impact of climatic variation on geographical distribution patterns and processes. This impact can be less apparent at the fine temporal and spatial scales that much of ecology is acquainted with, but it can result in a dilemma for biogeographers (e.g. Enright, 1976). Moreover, discussions in ecology often concern arguments for or against the operation of abiotic versus biotic factors, or those of biotic factors superimposed upon

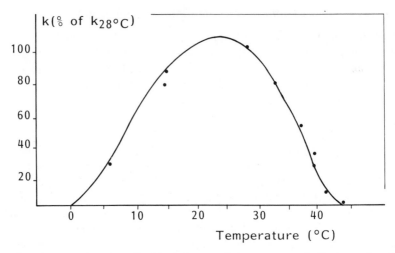

Fig. 10.1 Asymptotic values K of logistic growth in yeast populations as a function of temperature. Individual values are expressed as percentages of the value reached at 28°C (after Gause, 1932).

abiotic ones. They do not integrate them in the way Gause (1931, 1932) proposed, making abiotic factors conditional to effects of biotic ones. In his experiments, Gause found that K of the logistic growth equation follows a Gaussoid optimum response relationship to temperature (Fig. 10.1). Thus, temperature can be conditional to the intensity of logistic growth processes. This and similar relationships with other variables could explain the numerical build up of ranges in geographical space in terms of optimum response surfaces (Hengeveld and Haeck, 1981, 1982). The mechanism of such relationships can be varied; for abundance distributions of bumblebees in Kent, Williams (1988) suggested that energy economics are implied; other, independent studies point in the same direction (cf. also Currie and Paquin, 1987; Turner *et al.*, 1987; Wright, 1983).

For example, Kooijman (1988), applying Von Bertalanffy's growth equation to about 260 species of 9 phyla, described species size and life history in terms of energy budgets. Similarly, Koskimies and Lahti (1964) found that ducklings of ten species differ greatly in temperature sensitivity operating through basal metabolic rate, which relates to their plumage, life history, behaviour and geographical distribution. As to their geographical distribution, the most cold-hardy species are both the most widespread species in Europe as well as the most northern ones, whereas the reverse holds for the least cold-hardy species. Kendeigh (1976) described similar relationships for various parts in the range of the house sparrow in North America (cf. also Blem, 1973, 1980). Thus, these observations and experiments show that energy budgeting is important to many species and that it can condition the intensity of growth processes. Although the abiotic environment comprises more elements than just temperature, such as precipitation, these other elements can also affect the organism's energy budget through temperature, humidity or precipitation lowering the effective temperature (e.g. Pigott, 1970).

Instead of at the physiological level, Parry (1978) and Parry and Carter (1985) described the mechanism of the response of oats in Scotland to temperature at the level of populations, using a time-series of 323 years for estimating the annual heatsum of temperatures above 6°C. As the heatsum shows an altitudinal lapse rate, they estimated the uppermost altitude for oat growing in history, using the minimum annual sum for crop failure. Moreover, assuming the heatsum to be normally distributed in the period chosen, they constructed curves for the expected risk of yield failure for 1, 2, 3, . . . failures in succession. Temporal frequency distributions in intensity thus represent species' risk distributions. When there are trends in the time-series, the procedure is repeated for various parts of the series, thus for each part resulting in an altitudinal risk distribution (Fig. 10.2).

Wigley (1985) reasoned that extremes of, say, temperature are very sensitive to changes in the mean, being steeply curvilinear for more extreme events. As species are, moreover, sensitive to environmental variation at

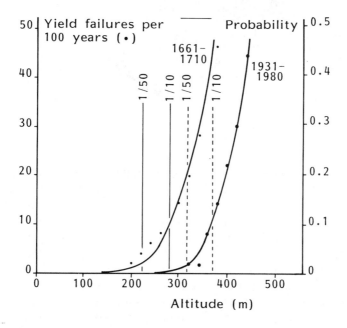

Fig. 10.2 Yield failure of oats as a function of altitude in Scotland for two historical periods. Dots indicate observed, drawn lines expected frequencies, assuming that frequency of failure at a certain altitude is normally distributed for the period concerned (after Parry and Carter, 1985).

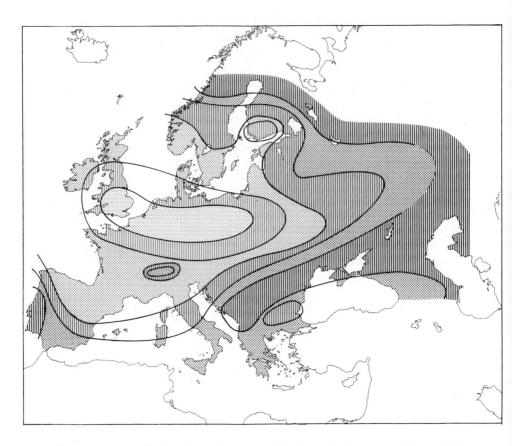

Fig. 10.3 Standard deviations of crop yield of cereals in western Europe (after Chisholm, 1980).

their range margins, the increase in variability towards the margins is understandable (Fig. 10.3) (cf. Klages, 1942; Uchijima, 1978). But species are not only most stable in their range centres, they are also most abundant there, occur in the largest aggregations, and in most habitats (Hengeveld and Haeck, 1981, 1982). Each of these characteristics can thus be considered a parameter of a characteristic at a higher level, the species' local intensity (Hengeveld and Haeck, 1981). Thus, towards their range margins, species are not confined to ecologically marginal habitats but are found in those that are most similar to optimum ones in their centres and hence to those with relatively the lowest frequency of failure. Invaders will therefore occupy optimal habitats first and from there spread out into the more risky, marginal ones.

Over time, the dome-shaped distribution of the species' expected intensities can shift, thus reflecting shifts in frequency of occurrence of climatic conditions. As local densities express intensities at certain distances from the range centre, local density fluctuations reflect geographical shifts in this dome-shaped distribution. Locally, these shifts are experienced as invasions, outbreaks, range contractions or extinctions.

Consequently, as responses to statistically varying climatic elements and stochastic spatial processes, abundance distributions are also statistical entities. Diffusion, accordingly, is just one of three parameters in the response process as a whole (known as a general-transport model), besides those of species abundance or intensity (the 'source/sink' parameter) and direction (the 'advection' parameter). When the three parameters representing these processes in geographical space are limiting, the range location will be stationary and the internal intensity distribution will reflect the climatic conditions. They then describe spatial climatic variability rather than that of the species. When, however, spatial processes are not limiting, the species' range location and intensity distribution follow climatic changes in space. For example, advection in the beech in eastern North America reflects a northward extension of warm, moist air masses, lateral diffusion spatial variability in storm frequency, position and intensity, and an overall increase in intensity level and general climatic amelioration (Dexter *et al.*, 1987).

Invasions thus open the way to broader understanding of ecological and biogeographical processes and through these to the integration of other population biological disciplines. The condition, however, is that their presently dominating paradigms be dropped, and that they be replaced by one paradigm accounting for spatial dynamics on several scales.

10.9 Invasions and measures of control and conservation

Often, large population size is felt to be conditional for maintaining sufficient genetic variation, although it is often unclear how large populations minimally should be. For example, the 200 million starlings at present occupying North America stem from about 60 individuals liberated in New York. Also, the many millions of muskrats ravaging central and western Europe stem from only five escapees near Prague. These are two of several examples from a large number of cases indicating that, depending on the species, populations with sometimes surprisingly low thresholds can be viable, and remain so for apparently long periods of time (e.g. Prentice, 1976 for an annual plant species).

The frequent occurrence of invasions also indicates that spatial dynamics is essential to population dynamics. In accordance with the neglect of spatial dynamics, authors such as Mayr (1951, 1963) or Stresemann and Nowak (1958) adopt a spatially static view when they assume genetic changes as potential explanations of species invasions, implying that without such changes nothing would have happened. But in many cases, if not all, this assumption is still unsupported (cf. Gray, 1986). On the contrary, species should be viewed as heterogeneous, highly dynamic aggregations of individuals, each responding to ever changing environmental conditions. Because these responses are often unimodal, species intensity distributions can be described as optimum-response surfaces with regions of high and low expected intensities, known as gradation and latent areas, respectively.

Using the concepts 'gradation' and 'latent' area, German foresters of the 1930s and 1940s (cf. Schwerdtfeger, 1968) indicated areas within the range with highest and lowest outbreak frequency. Similarly, Zadoks and Rijsdijk (1984) as epidemiologists, constructed an atlas of European cereal disease. Often this work is confined to geographically static ranges, rather than to expanding ones. As one of many exceptions, however, Cook (1925) estimated the potential gradation and latent areas of the alfalfa weevil during its expansion phase after its introduction into North America (Chapter 6). Similarly, Pimm and Bartell (1980) estimated the potential range of the fire ant in North America (Chapter 6) and indicated conditions of eventual range delimitation and more optimal occurrence. Both analyses were done for short periods, accordingly assuming that climate is constant. But when climatic trends or temporally favourable periods do occur, species can grow rapidly from endemic to outbreak proportions and disappear as soon as the environment becomes unsuitable again. Thus, Pschorn-Walcher (1954) noticed that during the warm 1930s, 1940s and early 1950s, many continental insect species in Central Europe reached outbreak proportions and sank back to their low, endemic levels again after the mid-1950s. During the first part of this period an institute was even founded in the Netherlands for investigating forest pests which have since almost or completely vanished. At present in control programmes it is, therefore, common practice to consider climatic conditions in relation to the species.

Although nature reserves may occasionally suffer from pests, nature conservationists usually have to try and prevent species from dying out. Two types of extinction can be recognized in this context; chance extinction due to reserve size, and extinction due to climatic trends. The first type concerns extinctions mainly due to environmental variability, which results in a negative correlation between reserve size and extinction chance (e.g. Newmark, 1987; Soulé, 1987). The second type concerns extinction due to the fact that the location of reserves is fixed although they occur in a spatially dynamic environment. This discrepancy causes species loss when these shift from the area and hence out of the reserve. When the vicinity of the reserve is unsuitable, the species is lost for a larger area, or perhaps forever, because of a lack of spatial buffering. This cause of extinction will grow in significance when the climate warms up due to atmospheric changes (Peters and Darling, 1985). Thus, nature reserves should be large enough or located so as to prevent either type of extinction occurring. Moreover, they should also be large and heterogeneous to allow large populations to build up, at first as bridgeheads to invasions and later as permanent parts of a species' range.

10.10 Conclusions

Both ecology and biogeography contain dominant paradigms that assume that spatio-temporal processes are stationary. The frequent occurrence of invasions indicates that these processes are not stationary either in time or in space, and thus constitute anomalies to both paradigms. Instead they fit into an alternative idea or paradigm that, in principle, could replace the other two as a new one. This potentially new paradigm concerns physiological regulation rather than demographic regulation of species intensities and is based on energy budgeting specific to each species. Thus, ranges as two-dimensional geographical response surfaces would shift with changes in climate, invasions being one of its expressions.

Eleven
Conclusions

Wave fronts can be expressed by continuous lines on broad-scale maps, and their progression often by steady rates. At finer scales, however, this rate can vary, depending on topographical and weather conditions as Schröpfer and Engstfelt (1983) showed for the muskrat in Germany. At still finer spatial scales distinctions were made, first between bridgeheads ahead of the wave front, then between sharply defined fronts versus broad progression zones and finally between continuous versus broken-up fronts consisting of independently propagating foci.

The broad-scale, steady wave progression can still be described deterministically, as can the continuous, sharply defined wave front itself. But at finer spatial and temporal scales, this description is replaced by one framed in stochastic terms, spatial transition probabilities defining the dispersion probability field.

At ecological scales, it is interesting to look for differences in processes occurring within the wave front and those behind this front in established populations. Parameters with different values are, for example, reproduction rate and the number and quality of habitats where the species is found. At first species are confined to optimal habitats, after which they start occupying more marginal habitats, thus filling in space. However, populations within the wave front still expand exponentially, whereas those behind this front grow logistically. Skellam's (1951) equation of biological diffusion replacing the exponential growth component of Fisher (1937) and Kendall (1948) by one of logistic growth is therefore unnecessary. This, of course, is not universal; there are cases with a spatial progression that allow that the logistic equation also applies for the front populations.

Invasions are only the tip of the iceberg, often exhibiting common processes of internal range dynamics; they show the spreading rate of species into regions where they were absent before. Diffusion as a causative process is not a special class, but happens where differences in concentration occur. In all instances it represents the general movement of particles, in biological diffusion being supplemented by ecological parameters.

Species invasions can exhibit normal internal range dynamics. Within ranges, spatial dynamics not only account for recolonization of areas, but also for the chances of the sexes to meet, for young individuals to establish and for finding food or hiding places. Too often spatial aspects of ecological processes have been ignored, being left virtually unstudied for decades, although epidemiology and much of population genetics and evolutionary theory have long been familiar with the idea. Yet in the past, many ecologists emphasized the significance of migration, colonization or seed dispersal, but this evidence may not have fitted into the dominating paradigms of ecology and biogeography.

The more we know about geographical species distributions, the more dynamic they appear to be. Thus, species ranges are not only spatially structured with highest intensities in their central parts, but these structures are dynamic as well. When taken together, the spatial and the temporal nonstationarity of ecological processes will prove important to future ecological and biogeographical theorizing. Because of this, these processes warrant attention from ecologists and biogeographers alike.

I hope that this book can contribute to a change in attitude by having introduced the more dynamic thinking of other disciplines of population biology into ecology and biogeography. I also hope that it may integrate the various population biological disciplines concerned into a single approach for the study of processes above the level of the individual.

References

Ammerman, A.J. and Cavalli-Sforza, L.L. (1971) Measuring the rate of spread of early farming in Europe. *Man*, **6**, 674–88.

Ammerman, A.J. and Cavalli-Sforza, L.L. (1984) *The Neolithic Transition and the Genetics of Populations in Europe*. Princeton University Press, Princeton.

Ås, S. (1984) To fly or not to fly? Colonization of Baltic islands by winged and wingless carabid beetles. *J. Biogeogr.*, **11**, 413–26.

Baker, H.G. and Stebbins, G.L. (eds) (1965) *The Genetics of Colonizing Species*. Academic Press, New York.

Blem, C.R. (1973) Geographic variation in the bioenergetics of the house sparrow. *Ornithol. Monogr.*, **14**, 96–121.

Blem, C.R. (1980) The energetics of migration, in *Animal Migration, Orientation, and Navigation*, S.A. Gauthreaux (ed), Academic Press, New York, pp. 175–224.

Bozkho, S.I. (1980) *Der Karmingimpel*. Ziemsen Verlag, Wittenberg-Lutherstadt.

Broadbent, S.R. and Kendall, D.G. (1953) The random walk of *Trichostrongylus retortaeformis*. *Biometrics*, **9**, 460–66.

Briggs, J.C. (1974) The operation of zoogeographic barriers. *Syst. Zool.*, **23**, 248–56.

Brown, J.H. (1984) On the relationship between abundance and distribution of species. *Am. Nat.*, **124**, 255–79.

Brown, L.A. (1981) *Innovation Diffusion*. Methuen, London.

Brown, R.G.B. (1970) Fulmar distribution: a Canadian perspective. *Ibis*, **112**, 44–51.

Butler, L. (1953) The nature of cycles in populations of Canadian mammals. *Can. J. Zool.*, **31**, 242–63.

Carter, R.N. and Prince, S.D. (1981) Epidemic models used to explain biogeographical distribution limits. *Nature*, **293**, 644–45.

Cassie, R.M. (1962) Frequency distribution models in the ecology of plankton and other organisms. *J. anim. Ecol.*, **31**, 65–92.

Caughly, G. (1970) Liberation, dispersal and distribution of himalayan thar (*Hemitragus jemlahicus*) in New Zealand. *New Zealand J. Sci.*, **13**, 200–239.

Cavalli-Sforza, L.L. (1958) Some data on the genetic structure of human populations. *X Int. Congr. Gen.*, **I**, 389–407.

Chisholm, M. (1980) The wealth of nations. *Trans. Inst. Brit. Geogr.*, **5**, 255–76.

Clarke, C.M.H. (1971) Liberations and dispersal of red deer in northern South Island districts. *N.Z.J. For. Sci.*, **1**, 194–207.

Cliff, A.D., Haggett, P., Ord, J.D. and Versey, G.R. (1981) *Spatial Diffusion.* Cambridge University Press, Cambridge.

Cook, W.C. (1925) The distribution of the alfalfa weevil (*Phytonomus posticus* Gyll). A study in physical ecology. *J. Agric. Res.*, **30**, 479–91.

Cramp, S., Bourne, W.R.P., and Saunders, D. (1974) *The Seabirds of Britain and Ireland.* Collins, London.

Crawley, M.J. (1986) What makes a community invasible? in *Colonization, Succession and Stability.* A.J. Gray, P.J. Edwards and M.J. Crawley (eds), Symp. Brit. Ecol. Soc., **26**, Blackwell, Oxford, pp. 429–53.

Currie, D.J. and Paquin, V. (1987) Large-scale biogeographical patterns of species richness of trees. *Nature*, **329**, 326–27.

Cushing, D.H. (1982) *Climate and Fisheries.* Academic Press, London.

Danell, K., (1978) Ecology of the muskrat in northern Sweden. Thesis, Umeå.

Davis, M.B. (1976) Pleistocene biogeography of temperate deciduous forests. *Geoscience and Man*, **13**, 13–26.

Davis, M.B. (1981) Quaternary history and the stability of forest communities, in *Forest Succession.* D.C. West, H.H. Shugart and D.B. Botkin (eds), Springer, New York, pp. 132–53.

Davis, M.B. (1986) Climatic instability, time lags, and community disequilibrium. in *Community Ecology*, J. Diamond and T.J. Case (eds), Harper and Row, New York, pp. 269–84.

DeBach, P. (1965) Some biological and ecological phenomena associated with colonizing entomophagous insects, in *The Genetics of Colonizing Species*, H.G. Baker and G.L. Stebbins (eds), Academic Press, New York, pp. 287–306.

De Lattin, G. (1957) Die Ausbreitungszentren der holarktischen Landtierwelt. *Verh. deutsch. zool. Ges.*, **1956**, 380–410.

Dexter, F., Banks, H.T. and Webb, T. (1987) Modelling Holocene changes in the location and abundance of beech populations in eastern North America. *Rev. Palaeobot. Palynol.*, **50**, 273–92.

Diekmann, O. (1979) Run for your life. A note on the asymptotic speed of propagation of an epidemic. *J. Diff. Equat.*, **33**, 58–73.

Dobson, A.P. and May, R.M. (1986) Patterns of invasions by pathogens and parasites, in *Ecology of Biological Invasions of North America and Hawaii*, H.A. Mooney and J.A. Drake (eds), Springer, New York, pp. 58–76.

Dott, H.E.M. (1986) The spread of the house sparrow *Passer domesticus* in Bolivia. *Ibis*, **128**, 132–37.

Drost, J. (1962) *The Migrations of Birds.* Heinemann, London.

Duviard, M. (1977) Migrations of *Dysdercus* spp. (Hemiptera: Pyrrhocoridae) related to movements of the Intertropical Convergence Zone in West Africa. *Bull. ent. Res.,* **67**, 185–204.

Ecke, D.H. (1954) An invasion of Norway rats in southwest Georgia. *J. Mammal.,* **35**, 521–25.

Egerton, F.N. (1973) Changing concepts of the balance of nature. *Quart. Rev. Biol.,* **48**, 322–50.

Eighmy, J.L. (1979) Logistic trends in southwest population growth, in *Transformations. Mathematical Approaches to Cultural Change,* C.Renfrew and K.L. Cooke (eds), Academic Press, New York, pp. 205–20.

Elseth, G.D. and Baumgardner, K.D. (1981) *Population Biology.* Van Nostrand, New York.

Elton, C.S. (1958) *The Ecology of Invasions by Animals and Plants.* Methuen, London.

Elton, C.S. and Nicholson, M. (1942) The ten-year cycle in numbers of the lynx in Canada. *J. anim. Ecol.,* **11**, 215–44.

Enright, J.T. (1976) Climate and population regulation, the biogeographer's dilemma. *Oecologia,* **24**, 295–310.

Erkamo, V. (1956) Untersuchungen uber die pflanzenbiologischen und einige andere Folgeerscheinungen der neuzeitlichen Klimaschwankung in Finnland. *Ann. Bot. Soc. Zool. Bot. Fenn. 'Vanamo',* **28**, 1–290.

Erwin, T.L. (1981) Taxon pulses, vicariance, and dispersal: an evolutionary synthesis illustrated by carabid beetles. in *Vicariance Biogeography: A Critique,* G. Nelson and D.E. Rosen (eds), Columbia University Press, New York, pp. 159–96.

Erwin, T.L. (1985) The taxon pulse: a general pattern of lineage radiation and extinction among carabid beetles, in *Taxonomy, Phylogeny and Zoogeography of Beetles and Ants,* G.E. Ball (ed), Junk, The Hague, pp. 437–72.

Firbas, F. (1949) *Spät- und nacheiszeitliche Waldgeschichte Mitteleuropas nordlich der Alpen.* I. Fischer, Jena.

Fisher, J. (1952) *The Fulmar.* Collins, London.

Fisher, J. (1953) The collared turtle dove in Europe. *Brit. Birds,* **56**, 153–81.

Fisher, J. (1966) The fulmar population of Britain and Ireland, 1959. *Bird Study,* **13**, 5–76.

Fisher, R.A. (1937) The wave of advance of advantageous genes. *Ann. Eugen.,* **7**, 355–69.

Fisher, R.A., Corbet, A.S. and Williams, C.B. (1943) The relation between the number of species and the number of individuals in a random sample of an animal population. *J. anim. Ecol.,* **12**, 42–58.

Flade, M., Franz, D. and Helbig, A. (1986) Die Ausbreitung der Beutelmeise (*Remiz pendulinus*) an ihrer nordwestlichen Verbreitungsgrenze bis 1985. *J. Ornithol.,* **127**, 261–87.

Ford, M.J. (1982) *The Changing Climate*. Allen and Unwin, London.

Frakes, L.A. (1979) *Climates Throughout Geologic Time*. Elsevier, Amsterdam.

Gause, G.F. (1931) The influence of ecological factors on the size of population. *Am. Nat.*, **65**, 70–6.

Gause, G.F. (1932) Ecology of populations. *Quart. Rev. Biol.*, **7**, 27–46.

Glutz von Blotzheim, U.N. and Bauer, K.M. (1980) *Handbuch der Vögel Mitteleuropas, Vol. 9*. Akademische Verlagsgesellschaft, Wiesbaden.

Gray, A.J. (1986) Do invading species have definable genetic characteristics? *Phil. Trans. R. Soc. Lond.*, **B314**, 655–74.

Grime, J.P. (1986) The circumstances and characteristics of spoil colonization within a local flora. *Phil. Trans. R. Soc. Lond.*, **B314**, 637–54.

Haftorn, S. (1971) *Norges Fugler*. Universitets Vorlaget, Oslo.

Hägerstrand, T. (1969) *Innovation Diffusion as a Spatial Process*. Chicago University Press, Chicago.

Haggett, P., Cliff, A.D. and Frey, A. (1977) *Locational Analysis in Human Geography*. Edward Arnold, London.

Handtke, K. and Mauersberger, G. (1977) Die Ausbreitung des Kuhreihers, *Bubulcus ibis* (L.). *Mitt. Zool. Mus. Berlin, Suppl. Bd. 3, Ann. Orn.*, **1**, 1–78.

Harrison, R.G. (1980) Dispersal polymorphisms in insects. *Ann. Rev. Ecol. Syst.*, **11**, 95–118.

Harvey, P.H. and Ralls, K. (1985) Homage to the null weasle, in *Evolution*, P.H. Harvey and M. Slatkin (eds), Cambridge University Press, Cambridge, pp. 155–71.

Hastings, N.A.J. and Peacock, J.B. (1975) *Statistical Distributions*. Butterworths, London.

Hedrick, P.W. (1984) *Population Biology*. Jones and Bartlett, Boston.

Heesterbeek, J.A.P. and Zadoks, J.C. (1986) Modelling pandemics of quarantine pests and diseases: problems and perspectives. *Crop Protection*, **6**, 211–21.

Hengeveld, R. (1985) Dynamics of Dutch ground beetle species during the twentieth century. *J. Biogeogr.*, **12**, 389–411.

Hengeveld, R. (1986) Theories on biological invasions. *Proc. Kon. Ned. Akad. Wet.*, **C90**, 45–9.

Hengeveld, R. (1988a) Mechanisms of biological invasions. *J. Biogeogr.*, **15**, 819–28.

Hengeveld, R. (1988b) Mayr's ecological species criterion. *Syst. Zool.*, **37**, 47–55.

Hengeveld, R. (1989a) Dynamic Biogeography, Cambridge University Press, Cambridge.

Hengeveld, R. (1989b) Caught in an ecological web. *Oikos*, **54**, 15–22.

Hengeveld, R. and Haeck, J. (1981) The distribution of abundance. II. Models and implications. *Proc. Kon. Ned. Akad. Wet.*, **C84**, 257–84.

Hengeveld, R. and Haeck, J. (1982) The distribution of abundance. I. Measurements. *J. Biogeogr.*, **9**, 303–16.

Hengeveld, R. and Hogeweg, P. (1979) Cluster analysis of the distribution patterns of Dutch carabid species (Col.), in *Multivariate Methods in Ecological Work*, L. Orloci, C.R. Rao and W.M. Stiteler (eds), Intern. Cooperative Publishing House, Fairland, Maryland, pp. 65–86.

Hengeveld, R. and Stam, A.J. (1978) Models explaining Fisher's log-series abundance curve. *Proc. Kon. Ned. Akad. Wet.*, **C81**, 415–27.

Hengeveld, R. and Van den Bosch, F. (in prep) Bird monitoring and invasions.

Hintikka, V. (1963) Ueber das Grossklima einiger Pflanzenareale, in zwei Klimakoordinatensystemen dargestellt. *Ann. Bot. Sci. Zool. Bot. Fenn. 'Vanamo'*, **34**, 1–65.

Hofstetter, F.B. (1960) Mögliche Faktoren der Ausbreitung von *Streptopelia d. decaocto* Friv. *Proc. XIIth Int. Ornithol. Congr., Helsinki, 1958*, 299–309.

Holloway. J.D. and Jardine, N. (1968) Two approaches to zoogeography: a study based on the distributions of butterflies, birds and bats in the Indo-Australian area. *Proc. Linn. Soc. Lond.*, **179**, 153–88.

Howden, H.F. (1985) Expansion and contraction cycles, endemism and area: the taxon cycle brought full circle, in *Taxonomy, Phylogeny and Zoogeography of Beetles and Ants*, G.E. Ball (ed), Junk, The Hague.

Hudson, R. (1965) The spread of the collared dove in Britain and Ireland. *Brit. Birds*, **58**, 105–39.

Hudson, R. (1972) Collared doves in Britain and Ireland during 1965–1970. *Brit. Birds*, **65**, 139–55.

Hultén, E. (1937) *Outline of the History of Arctic and Boreal Biota During the Quaternary Period*. Almquist, Stockholm.

Huntley, B. and Birks, H.J.B. (1983) *An Atlas of Past and Present Maps for Europe: 0–13000 Years Ago*. Cambridge University Press, Cambridge.

Hutchinson, G.E. (1978) *An Introduction to Population Ecology*. Yale University Press, New Haven.

Infantosi, A.F.C. (1986) Interpretation of case studies in two communicable diseases using pattern analysis techniques. PhD Thesis, London.

Ito, Y. and Miyashita, K. (1965) Studies on the dispersal of leaf- and planthoppers. III. An examination of the distance-dispersal rate curves. *Jap. J. Ecol.*, **15**, 85–9.

Järvinen, O. and Ulfstrand, S. (1980) Species turnover of a continental bird fauna: northern Europe, 1850–1970. *Oecologia*, **46**, 186–95.

Jensen, B. (1973) Movements of the Red Fox (*Vulpes vulpes* L.) in Denmark investigated by marking and recovery. *Danish Rev. Game Biol.*, **8**, 1–21.

Johnson, C.G. (1957) The distribution of insects in the air and the empirical relation of density to height. *J. anim. Ecol.*, **26**, 479–94.

Johnson, C.G., Taylor, L.R. and Southwood, T.R.E. (1962) High altitude migration of *Oscinella frit* L. (Diptera: Chloropidae). *J. anim. Ecol.*, **31**, 373–83.

Kaisila, J. (1962) Immigration and expansion der Lepidopteren in Finnland in den Jahren 1869–1960. *Acta Entomol. Fenn.*, **18**, 1–452.

Kendall, M.G. (1948) A form of wave propagation associated with the equation of heat conduction. *Proc. Cambridge Phil. Soc.*, **44**, 591–593.

Kendeigh, S.C. (1974) *Ecology.* Prentice Hall, Englewood Cliffs.

Kendeigh, S.C. (1976) Latitudinal trends in the metabolic adjustments of the house sparrow. *Ecol.*, **57**, 509–19.

Kessel, B. (1953) Distribution and migration of the European starling in North America. *Condor*, **55**, 49–67.

Kingsland, S. (1982) The refractory model: the logistic curve and the history of population ecology. *Quart. Rev. Biol.*, **57**, 29–52.

Kingsland, S. (1985) *Modeling Nature.* Chicago University Press, Chicago.

Klages, K.H.W. (1942) *Ecological Crop Geography.* Macmillan, New York.

Klomp, H., Van Montfoort, M.A.J. and Tammes, P.M.L. (1964) Sexual reproduction and underpopulation. *Arch. Neerl. Zool.*, **16**, 105–10.

Kooijman, S.A.L.M. (1988) The Von Bertalanffy growth rate as a function of physiological parameters, in *Mathematical Ecology*, T.G. Hallman, L.J. Gross, and S.A. Levin (eds), World Scientific, Singapore, pp. 3–45.

Koskimies, J. and Lahti, L. (1964) Cold-hardiness of the newly hatched young in relation to ecology and distribution in ten species of European ducks. *Auk*, **81**, 281–307.

Kuhn, T.S. (1962) *The Structure of Scientific Revolutions.* Columbia University Press, Chicago.

Lacey, W.S. (1957) A comparison of the spread of *Galinsoga parviflora* and *G. ciliata* in Britain, in *Progress in the Study of the British Flora*, J.E. Lousley (ed), London, pp. 109–15.

Lamb, H.H. (1977) *Climate. Present, Past and Future, Vol. 2. Climatic History and the Future.* Methuen, London.

Lande, R. (1976) Natural selection and random drift in phenotypic evolution. *Evolution*, **30**, 314–34.

Lawton, J.H. and Brown, K.C. (1986) The population and community ecology of invading species. *Phil. Trans. R. Soc. Lond.*, **B314**, 607–17.

Leston, D. (1957) Spread potential and the colonisation of islands. *Syst. Zool.*, **6**, 41–6.

Leys, H.N. (1964) Het voorkomen van de turkse tortel (*Streptopelia decaocto* (Friv.)) in Nederland. *Limosa*, **37**, 232–63.

Leys, H.N. (1967) The census of the collared turtle dove in the Netherlands. *X. Bull. Int. Counc. Bird Preserv.*, 147–54.

Lindroth, C.H. (1949) Die fennoskandischen Carabiden, Vol. 3. Allgemeiner Teil. *Göteborgs Kungl. Vet. o. Vitt. Samh. Handl.*, Ser. B, **4**, 1–911.

Lindroth, C.H. (1957) *The Faunal Connections Between Europe and North America.* Almquist and Wiksell, Stockholm.

Lubina, J.A. and Levin, S.A. (1988) The spread of a reinvading species: range expansion in the Californian sea otter. *Am. Nat.*, **131**, 526–43.

MacArthur, R.H. (1972) *Geographical Ecology.* Harper and Row, New York.

MacArthur, R.H. and Wilson, E.O. (1967) *The Theory of Island Biogeography.* Princeton University Press, Princeton.

Mack, R.N. (1981) Invasion in *Bromus tectorum* L. into western North America: an ecological chronicle. *Agro–Ecosystems*, **7**, 145–165.

Mack, R.N. (1985) Invading plants; their potential contribution to population biology, in *Studies in Plant Demography*, J. White (ed), Academic Press, London, pp. 127–42.

Mack, R.N. (1986) Alien plant invasion into the intermountain west; a case history, in *Ecology of Biological Invasions of North America and Hawaii*, H.A. Mooney and J.A. Drake (eds), Springer, New York, pp. 191–213.

Mayr, E. (1926) Die Ausbreitung des Girlitz (*Serinus canaria serinus* L.). *J. Ornithol.*, **74**, 571–671.

Mayr, E. (1951) Speciation in birds. *Proc. X Int. Ornithol. Congr. Uppsala, **1950***, 91–131.

Mayr, E. (1963) *Animal Species and Evolution.* Belknap Press, Cambridge, Mass.

Mayr, E. (1965) The nature of colonizations in birds. in *The Genetics of Colonizing Species*, H.G. Baker and G.L. Stebbins (eds), Academic Press, New York, pp. 29–47.

Mayr, E. (1982) *The Growth of Biological Thought.* Belknap Press, Cambridge, Mass.

Menozzi, P., Piazza, A. and Cavalli-Sforza, L.L. (1978) Synthetic maps of human gene frequencies in Europeans. *Science*, **201**, 786–92.

Mollison, D. (1977) Spatial contact models for ecological and epidemic spread. *J.R. Statist. Soc.*, **B39**, 283–326.

Mundinger, P.C. and Hope, S. (1982) Expansion of the winter range of the house finch: 1947–1979. *Amer. Birds*, **36**, 347–53.

Nelson, G. and Rosen, D.E. (eds), (1981) *Vicariance Biogeography. A Critique.* Columbia University Press, New York.

Newmark, W.D. (1987) A land-bridge island perspective on mammalian extinctions in western North American parks. *Nature*, **325**, 430–32.

Newton, I. (1985) *Finches.* Collins, London.

Niethammer, G. (1951) Arealveränderungen und Bestandschwankungen mitteleuropäischer Vögel. *Bonner zool. Beitr.*, **1951**, 17–54.

Nip-van der Voort, J., Hengeveld, R. and Haeck, J. (1977) Immigration rates of plant species in three Dutch polders. *J. Biogeogr.*, **6**, 301–8.

Nowak, E. (1975) *Ausbreitung der Tiere.* Ziemsen Verlag, Wittenberg Lutherstadt.

Okubo, A. (1980) *Diffusion and Ecological Problems: Mathematical Models.* Springer Verlag, Berlin.

Okubo, A. (in press) Diffusion-type models for avian range expansion. Symp. on approaches to biogeography, 19th Int. Ornithol. Congr., Ottawa, 1986.

Olsson, V. (1969) Die Expansion der Girlitzes (*Serinus serinus*) in Nordeuropa in den letzten Jahrzehnten. *Vogelwarte,* **25**, 147–56.

Olsson, V. (1971) Studies on less familiar birds. 165. Serin. *Brit. Birds,* **64**, 213–33.

Parry, M.L. (1978) *Climatic Change, Agriculture and Settlement.* Dawson and Archon, Folkstone.

Parry, M.L. and Carter, T.R. (1985) The effect of climatic variations on agricultural risk. *Climatic Change,* **7**, 95–110.

Peters, R.L. and Darling, J.D.S. (1985) The greenhouse effect and nature reserves. *Biosci.,* **35**, 707–17.

Pickett, S.T.A. (1980) Non-equilibrium coexistence of plants. *Bull. Torrey Bot. Club,* **107**, 238–48.

Pickett, S.T.A. and White P.S. (1985) *The Ecology of Natural Disturbance and Patch Dynamics.* Academic Press, Orlando.

Pielou, E.C. (1969) *An Introduction to Mathematical Ecology.* Wiley, New York.

Pielou, E.C. (1974) *Population and Community Ecology.* Wiley, New York.

Pielou, E.C. (1975) *Ecological Diversity.* Wiley, New York.

Pielou, E.C. (1977) *Mathematical Ecology.* 2nd ed. Wiley, New York.

Pielou, E.C. (1979) *Biogeography.* Wiley, New York.

Pigott, C.D. (1970) The response of plants to climate and climatic change, in *The flora of a changing Britain,* F. Perring (ed) p. 32–44. Hampton.

Pimm, S.L. and Bartell, D.P. (1980) Statistical model for predicting range expansion of the red imported fire ant, *Solenopsis invicta,* in Texas. *Environm. Entomol.,* **9**, 653–58.

Poole, R.W. (1974) *An Introduction to Quantitative Ecology.* McGraw-Hill, New York.

Prentice, H.C. (1976) A study in endemism: *Silene diclines. Biol. Conserv.,* **10**, 15–30.

Pschorn-Walcher, H. (1954) Die "Zunahme" der Schädlingsauftreten im lichte der rezenten Klimagestaltung. *Anz. Schädlingsk.,* **27**, 89–91.

Pyle, G.F. (1969) The diffusion of cholera in the United States in the nineteenth century. *Geogr. Anal.,* **1**, 59–75.

Rendine, S., Piazza, A. and Cavalli-Sforza, L.L. (1986) Simulation and separation by principal components of multiple demic expansions in Europe. *Am. Nat.,* **128**, 681–706.

Renfrew, C. (1987) *Archaeology and language.* Cambridge University Press, Cambridge.

Reynolds, J.C. (1985) Details of the geographic replacement of the red squirrel (*Sciurus vulgaris*) by the grey squirrel (*Sciurus carolinensis*) in eastern England. *J. anim. Ecol.,* **54**, 149–62.

Robbins, C.S. (1973) Introduction, spread, and present abundance of the house sparrow in North America. *Ornithol. Monogr.,* **14**, 3–9.

Robbins, C.S. (1988) *in press.*

Robbins, C.S., Bystrak, D. and Geissler, P.H. (1986) *The Breeding Bird Survey: its First Fifteen Years, 1965–1979.* US. Dept. Interior, Fish and Wildlife

Service, Resource Publication 157, Washington.

Rose, D.J.W. (1978) Epidemiology of maize streak disease. *Ann. Rev. Entomol.*, **23** 259–82.

Rosen, D.E. (1978) Vicariant patterns and historical explanation in biogeography. *Syst. Zool.*, **27**, 159–88.

Rucner, D. (1952) Die Cumra-Lachtaube in Jugoslawien. *Larus*, **4,5**, 56–73.

Safriel, U.N. and Ritte U. (1980) Criteria for the identification of potential colonizers. *Biol. J. Linn. Soc.*, **13**, 287–97.

Safriel, U.N. and Ritte, U. (1983) Universal correlates of colonizing ability, in *The Ecology of Animal Movement*. I.R. Swingland and P.J. Greenwood (eds), Oxford University Press, Oxford, pp. 215–39.

Salisbury, E.J. (1943) The flora of bombed areas. *North West Nat.*, **18**, 160–169.

Salomonsen, F. (1965) The geographical variation of the fulmar (*Fulmarus glacialis*) and the zones of marine environment in the North Atlantic. *Auk*, **82**, 327–355.

Sayers, B. McA., Mansourian, B.G., Phan Tan, T. and Bögel, K. (1977) A pattern-analysis study of a wild-life rabies epizootic. *Med. Inform.*, **2**, 11–34.

Sayers, B. McA., Ross, A.J., Saengcharoenrat, P. and Mansourian, B.G. (1985) Pattern analysis of the case occurrence of fox rabies in Europe in *Population Dynamics of Rabies in Wildlife*, P.J. Bacon (ed), Academic Press, London, pp. 235–54.

Schröpfer, R. and Engstfeld, C. (1983) Die Ausbreitung des Bisams (*Ondatra zibethicus* Linne, 1977, Rodentia, Arvicolidae) in der Bundesrepublik Deutschland. *Z. angew. Entomol.*, **70**, 13–37.

Schwerdtfeger, F. (1968) *Okologie der Tiere, Vol. 2. Demökologie.* Paul Parey, Hamburg.

Sharrock, J.T.R. (1976) *The Atlas of Breeding Birds in Britain and Ireland.* British Trust for Ornithology, Tring.

Shields, W.M. (1983) Optimal inbreeding and the evolution of philopatry, in *The Ecology of Animal Movement*, I.R. Swingland and P.J. Greenwood (eds). Oxford University Press, Oxford, pp. 132–59.

Siegfried, W.R. (1965) The status of the cattle egret in the Cape Province. *Ostrich*, **36**, 109–16.

Siegfried, W.R. (1966) The status of the cattle egret in South Africa with notes on neighbouring territories. *Ostrich*, **37**, 157–69.

Simberloff, D. (1981a) Community effects of introduced species, in *Biotic Crises in Ecological and Evolutionary Time*, M.H. Nitecki (ed), Columbia University Press, New York, pp. 53–81.

Simberloff, D. (1981b) What makes a good island colonist? in *Insect Life History Patterns; Habitat and Geographic Variation*, R.F. Denno and H. Dingle (eds), Springer, New York, pp. 195–205.

Simberloff, D. (1986) Introduced insects: a biogeographic and systematic perspective. in *Ecology of Biological Invasions of North America and Hawaii*, H.A. Mooney and J.A. Drake (eds), Springer, New York, pp. 3–26.

Simpson, G.G. (1977) Too many lines; the limits of the oriental and Australian zoogeographic regions. *Proc. Amer. Philos. Soc.*, **121**, 107–20.

Skellam, J.G. (1951) Random dispersal in theoretical populations. *Biometrika*, **38**, 196–218.

Skellam, J.G. (1952) Studies in statistical ecology. I. Spatial pattern. *Biometrika*, **39**, 346–62.

Skellam, J.G. (1955) The mathematical approach to population dynamics, in *The Numbers of Man and Animals*, J.B. Cragg and N.W. Pirie (eds), Oliver & Boyd, Edinburgh, pp. 31–46.

Skellam, J.C. (1973) The formulation and interpretation of mathematical models of diffusionary processes in population biology, in *The Mathematical Theory of the Dynamics of Biological Populations*, M.S. Bartlett and R.W. Hiorns (eds), Academic Press, London, pp. 63–86.

Smith, C.H. and Davis, J.M. (1981) A spatial analysis of wildlife's ten-year cycle. *J. Biogeogr.*, **8**, 27–35.

Soulé, M.E. (ed.) (1987) *Viable Populations for Conservation*. Cambridge University Press, Cambridge.

Southwood, T.R.E. (1962) Migration of terrestrial arthropods in relation to habitat. *Biol. Rev.*, **37**, 171–214.

Stjernberg, T. (1979) Breeding biology and population dynamics of the scarlet rosefinch *Carpodacus erythrinus*. *Acta Zool. Fenn.*, **157**, 1–88.

Stjernberg, T. (1985) Recent expansion of the scarlet rosefinch (*Carpodacus erythrinus*) in Europe. *Proc. XVIII Congr. Int. Ornithol.*, Moscow 1982, **2**, 743–53.

Stresemann, E. and Nowak, E. (1958) Die Ausbreitung der Turkentaube in Asien und Europa. *J. Ornithol.*, **99**, 243–96.

Thieme, H.R. (1977) A model for the spatial spread of an epidemic. *J. Math. Biol.*, **4**, 337–51.

Thieme, H.R. (1979a) Asymptotic estimates of the solutions of non-linear integral equations and asymptotic speeds for the spread of populations. *J. reine u. angew. Math.*, **306**, 94–121.

Thieme, H.R. (1979b) Density-dependent regulation of spatially distributed populations and their asymptotic speed of spread. *J. Math. Biol.*, **8**, 173–87.

Turner, J.R.G., Gatehouse, C.M. and Corey, C.A. (1987) Does solar energy control organic disversity? Butterflies, moths and the British climate. *Oikos*, **48**, 195–205.

Uchijima, Z. (1978) Long-term change and variability of air temperature above 10°C in relation to crop production, in *Climatic Change and Food Production*, K. Takahashi and M.M. Yoshiro (eds), Tokyo, pp. 217–29.

Udvardy, M.D.F. (1969) *Dynamic Zoogeography*. Van Nostrand Reinhold, New York.

Valentine, J.W. (1968) The evolution of ecological units above the population level. *J. Paleont.*, **42**, 253–67.

Van den Bosch, F., Hengeveld, R., Metz, J.A.J. and Verkaik, A.J. A new method for analysing animal range expansion (in prep.).

Van den Bosch, F., Zadoks, J.C. and Metz, J.A.J. (1988a) Focus expansion in plant disease. I. The constant rate of focus expansion. *Phytopathol.*, **78**, 54–8.

Van den Bosch, F., Zadoks, J.C. and Metz, J.A.J. (1988b) Focus expansion in plant disease. II. Realistic parameter-sparse models. *Phytopathol.*, **78**, 59–64.

Van den Bosch, F., Zadoks, J.C. and Metz, J.A.J. (1988c) Focus expansion in plant disease. III. Two experimental examples. *Phytopathol.*, (in press).

Van der Plank, J.E. (1963) *Plant Diseases; Epidemics and Control.* Academic Press, New York.

Van der Plank, J.E. (1967) Spread of plant pathogens in space and time, in *Airborne Microbes*, P.H. Gregory and J.L. Monteith (eds), Cambridge University Press, Cambridge, pp. 227–46.

Verkaar, H.J., Schenkeveld, A.J. and van de Klashorst, M.P. (1983) The ecology of short-lived forbs in chalk grasslands: dispersal of seeds. *New Phytol.*, **95**, 335–44.

Von Haartman, L. (1949) Der Trauerfliegenschapper. I. Ortstreue und Rassenbildung. *Act. Zool. Fenn.*, **56**, 1–104.

Von Haartman, L. (1973) Changes in the breeding bird fauna of North Europe, in *Breeding Biology of Birds*, Natl. Acad. Sciences. Washington DC., pp. 448–81.

Whittaker, R.H. (1967) Gradient analysis of vegetation. *Biol. Rev.*, **42**, 207–64.

Wigley, T.M.L. (1985) Impact of extreme events. *Nature*, **316**, 106–7.

Williams, E.J. (1961) The distribution of larvae of randomly moving insects. *Austr. J. Biol. Sci.*, **14**, 598–604.

Williams, P.H. (1988) Habitat use by bumblebees (*Bombus* spp.). *Ecol. Entomol.*, **13**, 223–37.

Wing, L. (1943) Spread of the starling and English sparrow. *Auk*, **60**, 74–87.

Wolfenbarger, D.O. (1975) *Factors Affecting Dispersal Distance of Small Organisms.* Exposition Press, New York.

Wright, D.H. (1983) Species–energy theory: an extension of species–area theory. *Oikos*, **41**, 496–506.

Wynne-Edwards, V.C. (1962) *Animal Dispersion in Relation to Social Behaviour.* Oliver and Boyd, Edinburgh.

Zadoks, J.C. and Kampmeier, P. (1977) The role of crop populations and their deployment, illustrated by means of a simulator, EPIMUL76. *Ann. New York Acad. Sci.*, **287**, 164–90.

Zadoks, J.C. and Rijsdijk, F.H. (1984) *Atlas of Cereal Diseases and Pests in Europe.* PUDOC, Wageningen.

Zadoks, J.C. and Schein, R.D. (1979) *Epidemiology and Plant Disease Management.* Oxford University Press, New York.

Species index

English

Alfalfa weevil 80
American robin 61

Blackbird 115

Canadian lynx 24
Carabid beetles 133
Cattle egret 112, 115
Cettis warbler 115
Cheat grass 51
Cholera 18, 49
Citrine wagtail 115
Collared dove 92ff.
Common grebe 115

Fire ant 81
Fulmar 103

Gallant soldier 47
Grey squirrel 127
Ground beetles 115, 133

Himalayan thar 17
House finch 112, 115

Japanese beetle 28

Kittiwake 115

linnet 107

Measles 19, 32
Muskrat 28, 65ff., 115

North American beech 9, 64, 137
Norwegian rat 127

Oak 48
Oats 130, 135

Penduline tit 103
Pied flycatcher 115
Potato blight 37

Rabies 117ff.
Red deer 14, 29, 115
Red squirrel 127
Reed bunting 115

Sand martin 115
Scarlet rosefinch 103
Serin 103
Shaggy soldier 47
Snowshoe hare 21
Starling (European) 39ff., 115

Turtle dove 102

Latin

Abies 9

Bromus tectorum 51
Bubulcus ibis 112, 115
    ibis 114
    coromandus 114

Carduelis canabina 107
Carpodacus erythrinus 103
    mexicanus 112, 115
Castanea dentata 9
Cervus elaphus 14
Cettia cetti 115

Emberiza schoeniculus 115

Fagus grandifolia 9
Fidecula hypoleuca 115
Fulmarus glacialis 103

Galinsoga ciliata 47
    parviflora 47

Hemithragus jemlahicus 17

Lepus americanus 21
Lynx canadensis 24

Monticilla citronella 115

Odontra zibethicus 65ff.

Phytonomus posticus 80
Phytophthora infestans 37
Picea spp. 9
Podiceps cristatus 115
Popillia japonica 28

Rattus norvegicus 127
    rattus 127
Remiz pendulinus 103
Riparia riparia 115
Rissa tridactyla 115

Sciurus carolinensis 127
    vulgaris 127
Serinus serinus 103
Solenopsis invicta 81
Streptopelia decaocto 86ff.
    turtur 102
Sturnus vulgaris 39, 115

Tsuga canadensis 9
Turdus merula 115
    migratorius 61

Author index

Ammerman, A.J. and Cavalli-
   Sforza, L.L. 13, 14, 62, 78, 79, 88
Ås, S. 55, 56

Baker, H.G. and Stebbins, G.L. 4
Blem, C.R. 134
Bozkho, S.I. 106
Broadbent, S.R. and Kendall,
   D.G. 89
Briggs, J.C. 132
Brown, J.H. 36
Brown, L.A. 45, 57
Brown, R.G.B. 109
Butler, L. 21, 22, 23

Carter, R.N. and Prince, S.D. 32
Cassie, R.M. 54
Caughley, G. 17
Cavalli-Sforza, L.L. 57, 58, 59
Chisholm, M. 136
Clarke, C.M.H. 14, 15, 16
Cliff, A.D., Haggett, P., Ord, J.D.
   and Versey, G.R. 18, 19, 20, 21,
   45, 49, 60, 62
Cook, W.C. 80, 138
Cramp, S., Bourne, W.R.P. and
   Saunders, D. 107, 109
Crawley, M.J. 129
Currie, D.J. and Paquin, V. 134
Cushing, D.H. 133

Danell, K. 66
Davis, D.E. 42
Davis, M.B. 9, 10, 130, 131
De Bach, P. 129
De Lattin, G. 132
Dexter, F., Banks, H.T. and Webb,
   T. 63, 64, 137
Diekmann, O. 77, 88
Dobson, A.P. and May, R.M. 41, 43
Dott, H.E.M. 112
Drost, J. 6
Duviard, M. 48

Ecke, D.H. 127
Egerton, F.N. 126
Eighmy, J.L. 35
Elseth, G.D. and Baumgardner,
   K.D. 53
Elton, C.S. 4, 6, 27, 28, 29, 39, 64,
   127, 131
   and Nicholson, M. 24
Enright, J.T. 133
Erkamo, V. 103, 133
Erwin, T.L. 132

Firbas, F. 131
Fisher, J. 94, 107, 108, 109
Fisher, R.A. 4, 73, 140
   Corbet, A.S. and Williams, C.B.
   55

Flade, M., Franz, D. and Helbig, A. 104, 105
Ford, M.J. 103
Frakes, L.A. 130

Gause, G.F. 36, 134
Glutz von Blotzheim, U.N. and Bauer, K.M. 103
Gray, A.J. 138
Grime, J.P. 4

Haftorn, S. 107
Hägerstrand, T. 45, 57
Haggett, P., Cliff, A.D. and Frey, A. 45, 49
Handtke, K. and Mauersberger, G. 112ff.
Harrison, R.G. 48
Harvey, P.H. and Ralls, K. 126
Hastings, N.A.J. and Peacock, J.B. 55
Hedrick, P.W. 126
Heesterbeek, J.A.P. and Zadoks, J.C. 5, 38
Hengeveld, R. 92ff., 103, 115, 126, 127, 130, 133
    and Haeck, J. 36, 62, 115, 130, 134, 136
    and Hogeweg, P. 62
    and Stam, A.J. 55
    and Van den Bosch, F. 92
Hofstetter, F.B. 102
Holloway, J.D. and Jardine, N. 132
Howden, H.F. 132
Hudson, R. 103
Hultén, E. 132
Huntley, B. and Birks, H.J.B. 9, 11, 12
Hutchinson, G.E. 97

Infantosi, A.F.C. 32, 33, 121ff.
Ito, Y. and Miyashita, K. 58

Järvinen, O. and Ulfstrand, S. 115

Jensen, B. 123
Johnson, C.G. 51
    Taylor, L.R. and Southwood, T.R.E. 52

Kaisila, J. 103, 133
Kendall, M.G. 140
Kendeigh, S.C. 41, 58, 61, 134
Kessel, B. 40
Kingsland, S. 36
Klages, K.H.W. 136
Klomp, H., Van Montfoort, M.A.J. and Tammes, P.M.L. 17
Kooijman, S.A.L.M. 134
Koskimies, J. and Lahti, L. 134
Kuhn, T.S. 126

Lacey, W.S. 47
Lamb, H.H. 103
Lande, R. 44
Lawton, J.H. and Brown, K.C. 127
Leston, D. 4
Leys, H.N. 95, 98
Lindroth, C.H. 127, 132
Lubina, J.A. and Levins, S.A. 115

MacArthur, R.H. and Wilson, E.O. 4, 53, 131, 132
Mack, R.N. 49ff.
Mayr, E. 103, 104, 127, 137
Menozzi, P., Piazza, A. and Cavalli-Sforza, L.L. 87
Mollison, D. 39, 46
Mundinger, P.C. and Hope, S. 110, 112

Nelson, G. and Rosen, D.E. 132
Newmark, W.D. 138
Newton, I. 106
Niethammer, G. 115
Nip-van der Voort, J., Hengeveld, R. and Haeck, J. 53
Nowak, E. 27

Okubo, A. 48, 101
Olsson, V. 103, 104

Parry, M.L. 130, 135
   and Carter, T.R. 130, 135
Peters, R.L. and Darling, J.D.S. 138
Pickett, S.T.A. and White, P.S. 131
Pielou, E.C. 36, 45, 46, 55
Pigott, C.D. 135
Pimm, S.L. and Bartell, D.P. 81, 83,
   84, 138
Poole, R.W. 36
Prentice, H.C. 137
Pschorn-Walcher, H. 138
Pyle, G.F. 18, 49, 62

Rendine, S., Piazza, A. and Cavalli-
   Sforza, L.L. 86ff.
Renfrew, C. 13. 45
Reynolds, J.C. 127
Robbins, C.S. 112
   Bystrak, D. and Geissler,
      P.H. 112, 133
Rose, D.J.W. 48
Rosen, D.E. 132
Rucner, D. 102

Safriel, U.N. and Ritte, U. 4
Salisbury, E.J. 48
Salomonsen, F. 109
Sayers, B. McA, Mansourian, B.G.,
   Phan Tan, T. and Bogel, K. 117
   Ross, A.J., Saengcharoenrat, P.
      and Mansourian, B.G. 117ff.
Schröpfer, R. and Engstfeld, C. 66,
   67, 140
Schwerdtfeger, F. 138
Sharrock, J.T.R. 109
Shields, W.M. 115
Siegfried, W.R. 112
Simberloff, D. 4, 41, 129

Simpson, G.G. 132
Skellam, J.G. 17, 18, 28, 48, 85, 88,
   90, 98, 102, 140
Smith, C.H. and Davis, J.M. 24
Soulé, M.E. 138
Southwood, T.R.E. 48
Stjernberg, T. 106, 115
Stresemann, E. and Nowak, E. 92,
   96, 103, 137

Thieme, H.R. 77, 88
Turner, J.R.G., Gatehouse, C.M. and
   Corey, C.A. 134

Uckijima, Z. 136
Udvardy, M.D.F. 132

Valentine, J.W. 132
Van den Bosch, F., Hengeveld, R.,
   Metz, J.A.J. and Verkaik, A.J. 92,
   101
   Zadoks, J.C. and Metz, J.A.J.
      88ff.
Van der Plank, J.E. 26, 38
Verkaar, H.J., Schenckeveld, A.J.
   and Van de Klashorst, M.P. 47,
   48
Von Haartman, L. 61, 115

Whittaker, R.H. 36
Wigley, T.M.L. 135
Williams, E.J. 89
Williams, P.H. 134
Wing, L. 40, 112
Wolfenbarger, D.O. 56, 60
Wright, D.H. 134
Wynne-Edwards, V.C. 109

Zadoks, J.C. and Kampmeier, P. 37
   and Rijsdijk, F.H. 138
   and Schein, R.D. 30

Subject index

Advection term 64, 137
Allee effect 17, 137
Areal circumference 27
  increase 28
  saturation 10, 97, 106

Balance-of-nature paradigm 126, 132
Barriers 16, 120
Bessel function 45, 89, 95
Blood groups 13
Bridgeheads 40, 95, 106, 112, 116

Carrying capacity 36
Catchment rate 89
Climographs 80
Colonization 7
Community adjustment 52, 102
  of invaders 131
Contact distribution 89
Culture follower 102, 114
Cyclic fluctuation 21
Cyclicity of epidemics 19

Deterministic processes 44
Diffusion 44, 45
  constant 61, 89
  hierarchical 18, 60
  long-distance 18, 46, 95
  neighbourhood 18, 46
  rate 61
  term 62, 137

Discontinous invasion 92, 106
Disharmonic biotas 53
Dispersal 7
  density 116
  jump 46
  long-distance 46, 95
  of sexes 16, 115
  short-distance 46, 95
  stratified 48, 107, 112
Dispersion 7
  probability field 56, 79, 95, 116, 117, 140
Distribution
  contact 88
  core 16
  geographical 7
  optimum response itensity 134, 137
  risk 135

Effects on invasions
  climate 81, 103, 107, 109
  focus number 51
  food conditions 102, 107, 109
  genetic constitution 103, 109 137
  natural enemies 41, 102, 107, 114
Endemism 19
Energy budgeting 134
Epidemic 7
Equilibrium theories 126ff.
Expansion
  bilateral 27

rate 79, 86, 95
 unilateral 27
 wave 87
Experimental analysis 48
Exponential growth 77, 95, 96, 101,
 107, 140
Extinction 32, 34

Focal trajectories 120
Foci 118, 119
Founder populations 51
Frequency distribution of
 population traits 45

Gene pool 23
General-transport model 62, 137
Genetic variation 137
 wave of advance 13, 60

Historical paradigm 133
Holocene invasions 9ff.

Information field 57, 116
 mean 57
Interception rate 89
Interval distribution of population
 traits 54
Invasion 7
 front 86
 open, closed fronts 46, 106
Irruption 7
Iso-probability maps 117

Linear growth 30
Logistic growth 13, 99, 134, 140

Mesolithic hunters 13, 86ff.
Micro-epidemics 32
Micro-epizootic profile 122
Migration 7
Minimum population size 17, 137
Monocyclic diseases 30

Neighbourhood area 57
Neolithic farmers 13, 86ff.

Nodal area 24
Nomadism 7
Non-stationarity 34

Peripheral zone 16
Polycyclic diseases 30
Population
 coalescence 9, 16
 traits
  frequency distribution 45
  interval distribution 45, 54
Primary gradient 116
Principal components analysis 81, 87
Probability distribution (probability
 density)
 Bessel function 45, 89, 95
 binomial 53
 double exponential 55
 exponential 45, 57, 89, 95
 Gaussian (normal) 45, 53
 log-normal 54
 log-Poisson 54
 log-series 55
 multinomial 53
 negative binomial 55
 Neyman-type A 55
 Poisson 54
 shifted-Gamma 89
Propagule disperson 48

Qualitative spread 21
Quantitative spread 21

Radial equivalence 17
 increase 26
Range expansion 7
 extension 7
Recurrent waves 21
Refugia 9
Regulation theory 129
Reproductive probability density 88
Risk distribution 135
 spreading theory 129
 theory 130

Rotational symmetry 57

Source/sink term 64, 137
Spatial expectation density 116
    saturation 97, 104, 106, 109, 140
    scales 37
    spread 7
Spectral analysis 24, 118
Square root of area 28
Stationarity 129
Steepness wave 89, 101

Stochastic processes 34, 44, 116ff.
Survival rate 77

Time kernel 88, 89
Transition probability 140

Wave of advance 86
    steepness slope 89, 101
    velocity 89, 99
Wing dimorphism 48, 56